AF888112

Comet Halley 1909-1911

COMET HALLEY 1909-1911

Retrospect of Observations

Edited by Diedrich Möhlmann
and Konrad Sauer

With 10 Figures and 39 Tables

Akademie-Verlag Berlin
1986

Erschienen im Akademie-Verlag Berlin, DDR-1086 Berlin, Leipziger Straße 3—4
© Akademie-Verlag Berlin 1986
Lizenznummer: 202 · 100/454/86
Printed in the German Democratic Republic
Gesamtherstellung: VEB Druckerei „Thomas Müntzer", 5820 Bad Langensalza
Lektor: Dipl.-Met. Heide Deutscher
Einbandgestaltung: Ralf Michaelis
LSV 1495
Bestellnummer: 763 558 0 (6933)

02500

Contents

Vorbemerkung 7
Introduction 10
September 1909 12
October 1909 14
November 1909 15
December 1909 17
January 1910 20
February 1910 21
March 1910 24
April 1910 26
May 1910 41
June 1910 132
July 1910 154
Last observations in 1910 154
Observations in 1911 156
Summary 158

Vorbemerkung

Das Erscheinen des Kometen Halley in Perihelnähe im Jahre 1986 und die zu seiner Erforschung vorgesehenen Experimente (VEGA, GIOTTO, Planet A und zum Teil auch ICE) sind der Anlaß für weltweit einsetzende Aktivitäten, die darauf gerichtet sind, aus den zu erwartenden Meßdaten eine maximale wissenschaftliche Ausbeute zu erzielen. Thematische Vorbereitungen dazu erstrecken sich über einen breiten Kreis von Problemen. Sie betreffen den Kometenkern und seine noch völlig unbekannte Oberfläche, seine Geschichte, die Größe, Struktur und chemische Zusammensetzung, die chemischen Prozesse in der Kometenatmosphäre und die Wechselwirkung des Sonnenwindes mit dem aus dem Kometenkern abströmenden Neutralgas und dem in ihm enthaltenen Staub.

Der Komet Halley ist eng mit der Geschichte der Kometenforschung verbunden, und seine Beobachtung läßt sich viele Jahrhunderte zurückverfolgen. Seit seiner letzten Wiederkehr im Jahre 1910 hat sich das Wissen über die Physik der Kometen beträchtlich erweitert. Entscheidend haben dazu neben den bodengebundenen astronomischen Methoden auch die neuen diagnostischen Möglichkeiten der Raumfahrt seit etwa 1970 beigetragen. Für die Erforschung speziell des Kometen Halley können aber auch die Beobachtungsergebnisse vom letzten Periheldurchgang mit genutzt werden, um theoretische Modelle aufzustellen, die helfen sollen, sich auf die vielfältigen neuen Beobachtungen vorzubereiten und diese tiefgründiger zu verstehen. Ein interessanter Punkt z. B. betrifft die beobachteten Kernteilungen, für deren Entstehung nach jüngsten Ergebnissen neben der Rotation des Kerns und lokalen Gasausbrüchen und resultierenden Brüchen auch Thermospannungen in Betracht gezogen werden müssen. Aber auch Erscheinungen wie starke Helligkeitsschwankungen und Anomalien in der Schweifstruktur sind von aktuellem Interesse.

Einige dieser Probleme werden auch im Institut für Kosmosforschung der Akademie der Wissenschaften der DDR bearbeitet. Die Durchsicht der verfügbaren Literatur über den Kometen Halley erfolgte zunächst nur unter dem Aspekt der eigenen Aufgaben. Erst später entstand der Gedanke, diese Literatur, die nur noch in wenigen Bibliotheken lückenlos vorhanden ist, in etwas aufbereiteter Form zu publizieren und damit einem großen Kreis von Interessenten zugänglich zu machen. Das Hauptanliegen besteht darin, durch chronologische Darstellung der unterschiedlichsten Beobachtungsergebnisse aus den Jahren 1909—1911 ein möglichst vollständiges Bild über das letzte Erscheinen des Kometen Halley zu vermitteln. Lediglich in einigen Fällen, wo Beobachtungen eines Autors über einen längeren Zeitpunkt mitgeteilt werden, erschien es zweckmäßig, von der zeitlichen Folge abzuweichen. Auf eine Wertung oder Kommentierung des historischen Materials wurde bewußt verzichtet. Die durchgehende zeitliche Anordnung wird in der obersten Kolumne einer jeden Seite durch die Angabe des jeweiligen Beobachtungsmonats kenntlich gemacht, um ein Auffinden spezieller Ereignisse zu erleichtern. Diese chronologische Anordnung machte auch ein thematisch orientierendes Inhaltsverzeichnis unmöglich und auch unnötig. Um dennoch schnell spezielle Phänomene, wie das gelegentliche „Aufblitzen" des Kerns, Ausbrüche, „jets" und sekundäre bzw. multiple Kerne auffinden zu können, sind die bemerkenswertesten Ereignisse, über die auf der jeweiligen Seite berichtet wird, stichwortartig ebenfalls als Kolumnentitel in der obersten Zeile einer jeden Seite angegeben. In der Tat hat gerade das Literaturstudium der Beobachtungen des Halleyschen Kometen anläßlich seines Periheldurchganges im Jahre 1910 Hinweise auf viele wieder vergessene bzw. nicht weiter beachtete Phänomene bei diesem Kometen gegeben, die für die Vorbereitung auf die intensiven Untersuchungen im Jahre 1986 und auf die Auswertung der erhaltenen Daten wesentlich sind. Hervorgehoben seien beispielsweise nur die Granulation der inneren Koma z. B. am 14. 12. 1909, die enorme Vielzahl von „jets", die nur als eine Folge offenbar sehr häufiger Ausbrüche zu verstehen sind und die in engem Zusammenhang mit häufigen Lichtwechseln zu sehen sind, z. B. auch als mehrfaches „Aufblitzen", wie am 8. Januar und am 23. Mai 1910 beobachtet, die zeitweilige „Zwiebelgewächs"-Struktur des Kometen, z. B. am 4. Februar 1910, sekundäre und multiple Kerne, die auf ein Abplatzen größerer Blöcke von der Oberfläche hinweisen, wie z. B. am 16. und 17. April und den Folgetagen im Jahre 1910, also in Perihelnähe, und mit weiteren entsprechenden starken tagelangen Aktivitätsphasen mit Nebenkernen, „jets" und Ausströmungen sowohl Anfang und Ende Mai als auch noch einmal Anfang Juni 1910, scheinbare Explosionen im Schweif (21. 4. 1910), Dunkelräume direkt hinter dem Kern (21. 4. 1910), der fragliche Durchgang der Erde durch einen Nebenschweif des Kometen (18./19. Mai 1910), die vielfachen dynamischen Prozesse in den Schweifen und auch die am 24. Mai 1910 in Leipzig beobachtete Sternbedeckung durch den Kometen. Fraglos wird sich vieles davon auch im Jahre 1986 wiederholen und die Bodenbeobachtungsprogramme und vielleicht auch die Raumflugmissionen beeinflussen.

Eine wichtige Quelle für das Literaturstudium bildete die zusammenfassende Darstellung von N. T. Bobrovnikoff (Publications of the Lick Observatory, Vol. XVII, Part II, University California, 1930). Daneben wurden Beiträge aus den folgenden Zeitschriften verwertet: Astronomische Nachrichten, Publications of the Astronomical and Astrophysical Society of America, Transvaal Observatory Circular (Union Observatory, South Africa), Monthly Notices of the Royal Astronomical Society.

Wir hoffen, daß das vorliegende Material einem großen Kreis von Wissenschaftlern bei der thematischen Vorbereitung auf die Wiederkehr des Kometen Halley und für die wissenschaftliche Auswertung der Beobachtungen im Jahre 1986 und darüber hinaus nützliche Dienste leisten möge.

Der thematisch und insbesondere naturwissenschaftlich orientierte Leser sei zum tieferen Verständnis der dargestellten Phänomene auf unser Buch „Kometen“ im Akademie-Verlag Berlin hingewiesen.

Diedrich Möhlmann
Konrad Sauer

Introduction

The 1986 perihel passage of periodic comet Halley and the related space research activities (VEGA, GIOTTO, Planet A and partially ICE) are a cause for increasing scientific work dealing with comets. These preparations for an intensive interpretation refer to a wide range of scientific problems. They are related to the yet unknown and never seen cometary nucleus, its surface, history, size, structure and chemical composition, to the chemical processes in cometary atmospheres and to the interaction between the solar wind and the neutral gas and dust expanding supersonically from the nucleus.

The comet Halley is very tightly connected with the history of scientific cometary research. Observations of this comet can be traced back for more then 21 centuries. But especially since it's last appearence in 1910 our knowledge about comets has increased drastically. This is caused as well by more improved ground based astronomical observations as by the new diagnostic tools, given by space research since about 1970. But especially for comet Halley, observations and results from investigations, based on its 1910 appearence, can be used as a basis for further and new models, interpretations and preparations for the interpretation of the 1986 results. There are very interesting properties of comet Halley, e.g. splitting of the nucleus and temporary secondary nuclei, jets, explosions, light variations etc. Obviously, they are of current interest.

Some of these problems are also of interest for the related scientific work in the Institut für Kosmosforschung der Akademie der Wissenschaften der DDR. The study of the literature on comet Halley in 1909—1911 was made originally under the viewpoints of this scientific work in our institute. It was mainly the enormous amount of unknown or foregotten properties and phenomena of comet Halley, which encouraged us to summarize all this literature and to present the essential observations and results of 1910 in a chronologically ordered form, to give a greater number of scientists and amateurs the possibility to see what happened from 1909 to 1911 with this interesting comet. This chronological scheme is supplied in some cases by a complete representation of observations, made by one author for a longer time.

It shall be noted here, that to find special phenomena or events, the month and the most important phenomena are mentioned by Keywords at the top of every page. Some of these remarkable phenomena shall be mentioned here: granulation of the inner coma (14. 12. 1909), an enormous number of jets which are related with earlier light variations (8. 1. 1910), a bulb-like structure of the coma (4. 2. 1910), secondary and multiple nuclei, indicating the separation of blocks from the surface of the nucleus (16. 4. 1910, 17. 10. 1910 and the following days) jets and outbursts in May and early June 1910, the expected crossing of the tail of the comet by the Earth at 18/19 May, 1910, an observed star occultation by the comet (24. 05. 1910) and the enormous world of dynamical processes in cometary atmosphere and tail. There can be no doubt that these phenomena will be of direct importance for the deeper interpretation of the 1986 results.

The most important source of this book was the summarizing work of N. T. Bobrovnikoff (Publications of the Lick Observatory, Vol. XVIII, part II, University of California, 1931). Important contributions came also from the Journals Astronomische Nachrichten, Publications of the Astronomical and Astrophysical Society of America, Transvaal Observatory Circular (Union Observatory, South Africa) and Monthly Notices of the Royal Astronomical Society.

We hope that the retrospect of observations of comet Halley in 1910 will be a help for all, dealing with this comet, and especially with the scientific interpretation of the 1986 data, related with this comet.

Some aspects of the physics of comets, which can be used as a basis for more detailed discussions of the above mentioned observations has been published in the book "Kometen" by the Akademie Verlag Berlin. The interested reader is referred to this book.

D. Möhlmann
K. Sauer

September, 1909

Auffindung des Halleyschen Kometen 1909 c.

Bei der Zentralstelle lief am 12. September mittags das folgende Telegramm ein:
„Komet Halley 11 September $14^h 7^m\!.3$ Königstuhl $\alpha = 6^h 18^m\!.2$ $\delta = +17°11'$ Gr. 16^m *Wolf*."

Der Ort schließt sich sehr nahe der Ephemeride in A. N. 433° an, welche mit den Elementen der unter dem Motto: „Isti mirantur Stellam" bei der Astron. Gesellschaft eingereichten Preisarbeit berechnet ist.

Kb.

Astron. Nachr.
Nr. 4355, S. 179, 1909

After Wolf's announcement it was found on a plate taken at Helwan on August 24, and a Greenwich on September 9. (Report)

The comet had been eagerly looked for by many observers. O. J. Lee[18] concluded from his failure to detect it with the 24-inch Yerkes reflector that it was certainly below the 17th magnitude in December 1908 and January 1909. Similar conclusions were reached by J. H. Metcalf[19] who used his 12-inch doublet, allowing for the motion of the comet, from March to December 1908.

The first plate of Halley's Comet on Mount Hamilton was obtained on September 12.937, 1909, C 2043. The nucleus measured 3″ to 4″ in diameter.

C 2044a, September 13.987, shows a nucleus of 3″, and a coma of 6″. It is reproduced in plate 10, Figure 1. C 2045 of September 14.979 shows only a faint tail.

The comet was first visually observed in this apparition by S. W. Burnham through the 40-inch Yerkes refractor, [20] on the morning of September 16, 1909. It was observed by E. E. Barnard on September 17. He describes it on this date as of magnitude $12^1/_2$ (estimated). The measured diameter was 7″.2. There was possibly a nucleus of the 15th magnitude. "It appeared as a fleck of light surrounded by a faint nebulosity for a few seconds. A little later it was quite easy to observe and was estimated to be 15th magnitude."

C 2054, September 22.964 and C 2054a, September 23,006, show a nucleus 2″ in diameter and a coma of 5″.

On September 26, Barnard found the comet much brighter. It "seemed to be about 14 or $14^1/_2$ magnitude at faintest." Its diameter was measured as 9.″5. It was brighter in the middle, but there was no definite nucleus.

These earliest observations show that even at discovery the comet had a well-marked condensation in the center, of not a nucleus strictly speaking. From Bernard's and Mount Hamilton observations there seem to have been fluctuations both in brightness and diameter. The degree of condensation toward the center also changed from night to night. This is the more remarkable as on September 14,

[18] Pop. Astr. **17**, 160, 1909.
[19] Pop. Astr. **17**, 440, 1909.
[20] Pop. Astr. **18**, 191, 1910.

N. T. Bobrovnikoff, Publ. of the Lick Observ., Vol. XVIII, pt. 2, p. 334, 1931.

Ninetieth Annual General Meeting

Comet c 1909. A return of Halley's Comet, first definitely recognised and announced by Dr. Max. Wolf on September 11, though the image was suspected on a plate taken on August 28. This suspicion has since been confirmed. On examination, proved that the comet was photographed at Greenwich on September 9, and at Helwân as early as August 24. While these results are eminently satisfactory, the credit of the discovery naturally remains with Dr. Max Wolf, who, by his early announcement, made the observation of the comet possible elsewhere. At Mount Hamilton the comet was photographed on September 12, but does not appear to have been observed visually in the 36-inch refractor till October 16. Early observations were secured at Yerkes. In November the comet became visible in telescopes of moderate size, but in this case, again, the increase in brilliancy has not been steady and uniform.

Dating from the Helwân observation, the comet became visible some 239 days before perihelion passage, which, it may be assumed, will occur on April 19.7 G.M.T., or about three days later than the date assigned by Messrs. Cowell and Crommelin, neccesitating at the time of discovery, a correction to the ephemeris of only $+24^s$ in R.A. and $-4'$ in declination. Considering the influences exercised by Jupiter on the orbit during the period 1835—1910, a period in which the perturbations were nearly at a maximum, the accuracy of the prediction bears eloquent testimony to the care which must have been exercised in their computation. It is difficult to express adequately the accuracy attained by Messrs. Cowell and Crommelin in this research; but since the date of perihelion passage depends upon the mean motion, and this again upon the mean distance, it may be pointed out that the correction needed to the semi-major axis of the ellipse is only about 130.000 miles, a quantity of quite the same order of magnitude as the uncertainty of the Earth's distance from the Sun. The success of their achievement is deserving of the warmest recognition, and has been universally acknowledged.

Spectroscopic observations began early. On October 22, after two hours' exposure, a faint spectrum was photographed at the Lick Observatory. From that date the physical observations have been continued with greater success as the comet increased in brightness. It is reported that Americal astronomers have appointed a committee to ensure the continuity of the observations on a systematic plan.

Monthly Notices of R.A.S LXX. (1910), p. 341.

October, 1909

C 2065, October 9.046 and C 2066, October 9.962, show a nucleus of 3.″0 and a coma of 5.″4. C 2069, October 10.923 and C 2070, October 10.962, show only a trail, probably due to the central condensation of diameter 2.″6.

On October 12 at Nice, Javelle[21] observed the comet as approximately round, diameter from 10″ to 12″; sometimes a central condensation of 15th magnitude.

C 2072, October 12.910, shows a nucleus of 3.″5, coma 5.′8. On C 2073, October 12.935, the image is fuzzy.

On C 2084, October 14.000, the nucleus measures 2.″3, and the coma 4.″7. It began to assume an asymmetrical shape. C 2085 and C 2086 of the same date show the same effect.

The comet on C 2091, October 15.010, has a nucleus 2.″5 in diameter. The coma is nearly circular. Its diameter is 5.″0. The nucleus is on side. The direction of the axis of symmetry, C 2090, October 14.980, shows the same aspect. Barnard, on October 17, measured the diameter of the comet as 10″. Javelle, at Nice, on October 18 assigned the diameter of the nebulosity 15″ to 20″. The comet increased in brightness very considerably.

On October 19, Barnard measured the diameter of the comet as 15″. He estimated it to he 13th magnitude. The sky was very thick. This may mean that the comet was even brighter than 13th magnitude. It was brighter in the middle "to an indefinite fleck of light." On October 19 and 20, Javelle found the nebulosity from 10″ to 15″ in diameter with a stellar nucleus fo 14th to 15th magnitude.

C 2097, October 21.841, shows the nucleus only, diameter 2.″3. On C 2100, October 22.867, the nucleus is very small and sharp, 1.″5. The coma is 3.″1 in diameter. Objective prism spectrogram[22] W 179, October 22.969, shows faint images of CN IV and C IV almost disappearing on the strong continuous background. There is a decided central condensation corresponding probably to the nucleus.

November, 1909

Javelle measured the diameter of the coma on November 4, 6, 8, and 9 at 15″. Giacobini[23], at Paris on November 5, found the comet 5″ to 6″ in diameter and the nucleus of 14th magnitude.

On November 14, Barnard estimated the magnitude of the comet at 13^{m} "much brighter in the middle to almost a nucleus. During moments of steadiness there was apparently a minute point of light in the nebulosity. The comet was round with no signs of a tail. The measured diameter was 11.′6."

On C 2123, November 14.785, the nucleus measured 3.″5, and the coma 5.″0. C 2131, November 15.980, has several images because of the defective guiding. The images are of various intensities and give a good illustration of how much the diameter of the comet depends on exposure. The nucleus measured 3.″8 but there is another much smaller condensation of diameter 1.″9. Nebulosity extends 8″—10″.

C 2132, November 17.831. Nucleus 3″—5″, coma 15″.

The comet was observed at Oxford[24] on November 17 and 18. On November 17 the comet appeard "dull with a central nucleus occasionally brightening up. Suspect an elongation in p.a. 135°." On November 18 it was a "nebulous haze, 12″ in diameter." The nucleus preceded the center and was south of it.

[21] Observations by Javelle were made with the 76-cm refractor of the Nice Observatory. General reference for his observations; Bull. Astr. 27, 456, 1910.

[22] See note by W. H. Wright, Lick Obs. Bull. 5, 146, 1909.

[23] Equatorial 38-cm. aperture. Bull. Astr. 27, 394, 1910.

[24] 18-inch equatorial, Radcliffe Observatory, Monthly Notices 70, 497, 1910, Observations from Nov. 5, 1909, to Febr. 11, 1910.

N. T. Bobrovnikoff, Publ. of the Lick Obser., Vol. X VII, pt. 2, p. 334/335, 1931.

Nov. 7. During the second set of observations a faint star followed the comet closely. Possibly during the first set of observations it blotted the comet's nucleus out, and was observed instead of the comet.

H. A. Howe, Astron. Nachr. Nr. 4784, S. 132.

Nachdem noch am 5. November 1909 vergebens nach dem Kometen gesucht worden war, konnte er hier zuerst am 20. Nov. beobachtet werden. Sein Aussehen blieb unscheinbar bis zu seinem ersten Verschwinden in den Sonnenstrahlen am Abendhimmel. Er erschien als rundliches Objekt mit sehr schwachen äußeren Nebenpartien; in der Mitte hob sich nicht selten eine hellere, gegen die Umgebung schlecht begrenzte Kernscheibe ab, in der einigemale ein scharfes Kernchen gesehen wurde. Auch eine Granulation dieser inneren Partien von wechselnder Feinheit fiel mehrfach auf. Der Kometenkörper schien bisweilen leicht gedehnt in den in der nachfolgenden Übersicht notierten PW. Die physischen Beobachtungen sind am großen Refraktor bei Vergr. 154 und 207 angestellt.

1909—10	M.Z. Straßb.	Durchmesser	Gedehnt im PW	Kernhelligkeit	Bemerkungen
Nov. 20	10^h	1′.3	130°		Helligkeit am gr. Refr. geschätzt; grob granuliert
Dez. 13	9.8	2.5	70		Kernscheibe 0′.7 Drm., Helligkeit im gr. Refr. = $9^m.6$
14	6.8	1	95		
15	6.6	2			Kernscheibe 0′.7 Dm.
Jan. 29	8.8	2.3		$11^m.3$	Helligkeit im gr. Refr. = $10^m.2$
31	8.9	2.5			Durchmesser im Sucher = 7′; fein granuliert
Febr. 1	8.8	2.0			Durchmesser im Sucher = 6′
2	8.7	1.5		12	
10	6.8	2		11.0	
14	6.6	2.3		11.8	
16	7.0	2	60		
22	6.8	2.0			
28	6.9	1.3			Durchmesser im Sucher = 2′.8
März 4	7.0				
5	7.1	1			
7	7.1				
9	7.3				Durchmesser im Sucher = 3′

C. Wirtz, Astron. Nachr. Bd. 186 Nr. 4446, S. 92—94, 1911.

. . . This is an interesting indication of the granulation of the inner part of the coma. It is confirmed by observations of the same period by Charlois.[26] He observed on December 14 at Nice that "le noyau, par moments, parait granulé."

On November 30, Barnard measured the diameter of the comet as 41″. "The brightest portion was almost 7″ in diameter; and what was perhaps an ill-defined nucleus, was something like one or two seconds in diameter."

The interesting feature of these early observations is that diameter of the comet estimated or measured visually was considerably greater than its photographic diameter. The latter, as we have seen, depends very much on the exposure, but even so the visual diameter seems to be too large. This seems rather strange as the chief constituent of the envelopes at this time was cyanogen, more active photographically than visually. The large visual diameters are probably due to the action of the Swan bands C III and C II in the green and yellow which would not affect the ordinary phopographic plate.

The heliocentric distance of the comet was 3.0 A.U. on October 6 and 2.4 on December 6. In spite of this large distance the comet showed considerable activity; for example the asymmetric development of the coma and changes in brightness, with a decided brightening[27] on November 22.

December, 1909

On December 2 Javelle observed the comet to be 30″ to 40″ in diameter: there was diffused light for about 2′ around the comet.

Deslandres and Bernard[28] obtained photographs of the comet on December 6, 8, 13. The nucleus on December 6 was round with two faint curved streamers "like antennae of an insect" in the general direction of the prolonged radius vector. On December 7 they disappeared. The spectrograms (22° ultra-violet objective prism) showed distinct condensations at λ 4000 and λ 4100.

C 2137, 1909, December 13.796 (reproduced in Pl. 10, fig. 2), is the first photograph to show traces of a tail, in p.a. 83°, that is, nearly in the prolonged radius vector (p.a. 72°). The tail is 1.′5 long. C 2136, December 13.749, shows a bright central part 9″ in diameter. The coma ist 30″ in diameter.

The central part of the comet was observed by Wirtz, on December 13, to be elongated in p.a. 70° and on December 14, in p.a. 95° which roughly agrees with my measurement. Visual observations by Aitken[29] with the Lick 36-inch refractor on December 14, 15, 17, 18 indicates that the comet "appeared to be nearly round, the coma being 1′ ± in diameter with a sharp nucleus. The halo around the comet extended over an area nearly 15′ in diameter and was decidedly brighter and somewhat more extended on the following side."

C 2139, December 14.769, has double images due to the defect in guiding. The tail extends to 3′; the nucleus measures 10″; the brighter part of the coma 1′; the fainter part 6′.

C 2140, December 15.741. The tail extends 2′—3′ in the same p.a. previously. The images are double. C 2141, December 16.742, gives a trail of the comet. The nucleus is 4″ in diameter and the central part of the coma 20″.

On December 19, the first of the series of spectrograms by Parkhurst, was taken at the Yerkes Obersatory. It showed[30] a continuous spectrum with the maximum of intensity at λ 4000. The usual emissions CN IV and C IV were present but were not conspicuous on the bright background. CN IV was considerably brighter than C IV.

Barnard photographed the comet with the Bruce telescope on December 29 and found a slender and straight tail of 10′ length in p.a. 69°, that is, almost in the direction of the prolonged radius vector, p.a. 70°.

[26] Bull. Astr. 27, 153, 1910, 39-cm. equatorial.
[27] Philips, Monthly Notices, 70, 183, 1909.
[28] Compt. Rend. 149, 1101, 1909.
[29] Publ. Astr. Soc. Pac. 21, 259, 1909.
[30] See may article in Astroph. Jour. 66, 145, 1927.

N. T. Bobrovnikoff, Publ. of the Lick Observatory, Vol. X VII, pt. 2, p. 335, 336, 1931.

1909 December 10. La comête, estimée de 12ᵉ grandeur, apparaît ronde, sur 15″ à 20″ de diamètre, avec, au centre, une légère condensation. — Déc. 30. La comète, de 11ᵉ grandeur, a une tête de 30″ de diamètre, un peu allongée.

P. Chofardet, Astron. Nachr. Bd. 188, Nr. 4489, S. 3, 1911.

Note on Variations in Brightness of Halley's Comet

By Rev. *T. E. R. Philips. M.A.*

Since 1909 November 16 I have observed Halley's Comet with a $12^1/_2$-inch Calver reflector on every available occasion, and the observations show some considerable variations in the comet's brightness. The aspect of so delicate an object is of course specially liable to be affected by atmospheric conditions, but the observations which have been made on ten nights seem sufficiently numerous to practically eliminate error arising from this source. I have commonly adopted the method of limiting apertures in comparing results, and have allowed for the atmospheric conditions prevailing at the time.

The comet appeared unquestionably brightest on the night of November 22, when it could be seen with a stop $3^1/_2$ inches in diameter placed eccentrically so as to avoid the obstruction of the flat. The Moon was nearly ten days old at the time and the atmosphere rather hazy, but the comet was surprisingly brighter than on the previous night, when the conditions were more favourable.

Mr. Hollis, on the same night (November 22), using the 10-inch guiding telescope of the Astrographic Equatorial at Greenwich, estimated the brightness as about magnitude 10. The following night the comet was very inconspicuous with the full aperture of the reflector, and the conditions were not so inferior to those of the 22nd as to be held abbountable for the whole change. The comet was not seen again till the end of the month, but since then it has been repeatedly oberved under quite satisfactory conditions without regaining the brightness of November 22. On December 3 and 8 it could just be glimpsed on a dark sky with $3^1/_2$ inches of aperture, but on December 6, when the sky was very clear and transparent after rain and the stars brillant, it was quite invisible with a $4^1/_2$-inch eccentric stop. It was then probably as faint as 11'5 or 12 magnitude. On December 8 the comet was brighter again, and was estimated to be about 10'5 or 11 magnitude.

The fact that the comet was most easily seen under the inferior conditions of November 22, and especially difficult on December 6 when circumstances were apparently excellent, may be regarded as proving the reality of changes in its visual brightness.

The distance between the Earth and the comet has been steadily diminishing, though less rapidly than formerly, so that the unexpected brightness on November 22 was probably in the nature of a spurt. It has, however, been generally brighter than the magnitude based on its early photographic images, so its recent comparative faintness need not be held to indicate a feeble display near perihelion. The spurt on November 22 suggests that the reverse is possible, if not probable.

Ashtead: 1909 December 9.

Monthly Notices of R.A.S LXX, 2 (1910) 183.

January, 1910

Janv. 4. La comète se présente comme une nébulosité de 3′ de diamètre, avec noyau peu brillant, dont l'éclat est èvalué de 11e grandeur. Pas d'apparence de queue.

H. Morize, Astron. Nachr., Bd. 186, Nr. 4443, S. 35, 1911.

Noyau assez net. — 1910 Janv. 5. La comète, de 10^{e}à 11e grandeur, a une tête de 40″ de diamètre et un noyau central, d'aspect changeant. — Janv. 8. La comète semble plus étalée, le noyau est décentré.

P. Chofardet, Astron. Nachr. Bd. 188, Nr. 4489, S. 3, 1911.

C 2145, January 7.692. The nucleus is round and large, about 8″, but there is a central condensation much smaller, 3″—5″ in diameter (better seen on C 2146, January 7.742). There is a broad fan with the edges at p.a. 10° and 65°, that is, with the axis at p.a. 40°. It is about 40″ long. There is a narrow jet at p.a. 202° about 20″ in length. General faint nebulosity may be 1'.5 in diameter.

H. C. Wilson[31], with the 16-inch refractor of the Goodsell Obervatory, found on January 14 a nucleus "fairly sharp". On January 20 "there was no nucleus but pretty sharp condensation." Remarkable fluctuations in brightness of the comet were observed by Philips[32] on January 10.

Plates WS1, WS2, WS3, WS4 of January 28, 29, 30, Februar 1, show about 1° of a narrow structureless tail with a coma 3′—4′ in diameter.

[31] Astr. Jour. 26, 75, 1910.
[32] Jour. Brit. Astr. Assoc. 20, 309, 1910.

N. T. Bobrovnikoff, Publ. of the Lick Observ., Vol. X VII, pt. 2, p. 336, 337, 1931

1910 Genn. 4. Nucleo 10^{m}0 (B.). — Genn. 6. Nucleo 10^{m}8; testa 0'.7 (M). — Genn. 10. Nucleo 10^{m}6 (Z). — Genn. 11. Nucleo 10^{m}8 (M). — Genn. 15. Nucleo 9^{m}5; testa 0'.7, dissimmetrica, espandesi a SE (Z). — Genn. 27. Nucleo 9^{m}5—10^{m}0; testa 1'.5; astro visible al cercatore de 5 cm (Z).

E. Millosevich, Astron. Nachr., Bd. 186, Nr. 4456, S. 263, 1911.

February, 1910

February 4.694, we have the first unmistakable appearance of "Zwiebelgewächs" (bulb-like shape of the comet)[34]. The nucleus is sharp but slightly trailed, 3″—4″ in diameter. There is a bright circular envelope of 20″ diameter around the nucleus and one much larger and fainter, 2.′7 in diameter. One might suspect a still larger circular envelope of some 6′ diameter. The tail is very slender and extends for 8′ in p.a. 71° (p.a. of prolonged radius vector was 66°). There is a broad fan between p.a. 2° and 84° of 2.′7 length; it is embedded in the coma. Fainter emissions in p.a. 134° and 176° are of the same length.

WF2, February 4.694, has a tail of 1.′7 length.

Objective-prism spectrogram W 190, February 4.694, has CN IV considerably brighter than C IV. "Violet" spectrum. Central core of CN IV is 1.′3 in diameter, outer 4.′4. They correspond to the inner and outer envelopes on the direct photographs. These envelopes are always larger on the objective prism spectrograms, due to defective guiding. The spectrum is partially mixed with that of a star. There is no outer envelope at C IV. Other knots, especially C + H, belong to the nucleus only . . .

February 5.689, . . . The nucleus is sharp and bright, 7″ to 8″ in diameter, slightly elongated due probably to insufficient guiding. The tail is slender. It extends in p.a. 70° (that is, forming an angle of 3° with the prolonged radius vector) for 5.′5, very rapidly diminishing in brightness. At 3.′0 from the nucleus it has a kink (recognizable in the reproduction) which shows the motion of matter along the tail. This activity is interesting in view of the fact that the heliocentric distance was 1.5 A.U. There are several concentric envelopes gradually diminishing in brightness with recession from the nucleus. The innermost brightest envelope has a diameter 1.′5; two emission fans at p.a. 45° and 145° are ambedded in it. There is a general glow in p.a. 45° extending for 3.′1. This is probably the emission fan of the date before, the edges of which were in p.a. 2° and 84°. If so, it has increased in length from 2.′7 to 3.′1. Another glow of the same dimensions is in p.a. 145°. There is a rather definite jet in p.a. 70° from which the tail begins. The difference in the intensity between this jet and the tail is quite marked.

A tail of a different kind is seen on C 2156, February 10.685. Several fine streamers (visible on the reproduction) are generated from a jet in p.a. 61°. The length of the jet is 80″. There are at least three streamers issuing from its tip. There is also a short emission fan in p.a. 186°, about 23″ long, but the general glow is more uniform than on February 5. Its diameter is about 2.′7. There seems to be general emission on both sides of the jet for about 30°, especially in p.a. 91°. The streamers extend 7′—8′ from the nucleus. Faint condensations are recognizable in them.

The nucleus is small and bright, 3″ to 4″ in diameter. The central condensation is about 20″ in diameter. . . .

Objective-prism spectrogram W 193, February 10.685, shows distinctly only CN IV, C IV and C III.

C 2157, February 11.665, shows a tail consisting of several streamers. Its p.a. is 63° and it extends 10′ from the nucleus. It is again formed by a jet in p.a. 73° and 60″ long. It is difficult to say whether the change in p.a. and length of the jet as compared with February 10 is real or not. Quite possibly this is another jet formed overnight. There is strong emission 48″ long, in p.a. 173°. It is probably the same as observed the night before in p.a. 186°, which was then 23″ long. Although the measurement is difficult the changes seem to be real. Central condensation is 5″ in diameter. The general halo is about 80″ in diameter and much brighter in the direction of the tail. WF 5 shows 40′ of tail.

Objective prism spectrogram W 193a, of the same date, shows distinctly only CN IV and C IV. CN IV has a jet on the side of the tail which is absent from C IV. Its length is about 1′ which corresponds to the measurement from C 2157. No doubt the jet consisted chiefly of CN. The outer halo can be distinguished only in C IV.

At Vienna, Jaschke[35] observed the nucleus flashing up on February 19. A similar observation[36] was made by him on January 8: "Der Komet . . . hat deutlichen Kern 10^{m} bis $10^{\mathrm{m}}.5$, der bisweilen noch heller aufblitzt." On February 21 he says that "bisweilen macht es den Eindruck, als ob nicht ein Kern sondern mehrere verschiedene Helle vorhanden wären." As we have seen, this multiplicity of the cometary nuclei was noticed earlier by Wirtz and by Charlois.

On C2159, February 28.690, p.a. of the tail is 70° (p.a. of the prolonged radius vector was 61°). The tail represents a wedge of about 4°; its southern part is appreciably brighter. It can be followed for about 20′. As on earlier dates it issues from a jet in the same p.a. about 90″ long, but the break in continuity is not as striking as before. There are several fine streamers in the tail and broad emissions for about 40° on each side of the tail. They are of the same lengths as the jet but much fainter. There is a very striking short emission of 23″ in pa.a. 229°. The nucleus is 5″ in diameter and very bright. An envelope of 12″ diameter differs only slightly from the nucleus in intensity. The next circular envelope is 38″ in diameter and the outer one is hardly visible.

Objective prism plate "194a, same date, shows CN IV stronger and larger than C IV. The jet at p.a. 229° is well visible in CN IV and is 50″ ong. There is no such jet in C IV. The broad emission on the tail side is also visible in CN IV (length $1'.5$). In C IV it is much fainter. The C + H knots are strong . . .

[34] For the meaning of his term see my paper Astroph. Jour. 66, 439, 1927 and Jaegermann, Bredichin's Untersuchungen, p. 173.

[35] Astr. Nach. 188, 81, 1911.

[36] It is difficult to pronounce judgment on the validity of such observations. Rapid fluctuations in the light of the nucleus have been noticed by many experienced observers. These observations are so persistent that it seems unwise to neglect them altogether. It is likely that many observers would hesitate to publish such data, assuming them to be of psychological origin. Merfield (Journ. Brit. Astr. Assoc. 20, 201, 1910), also speaks of "a condensation which occasionally sparkles".

N. T. Bobrovnikoff, Publ. of the Lick Observatory, Vol. XVII, pt 2, p 338, 1931.

Place of Halley's Comet, 1909—1910, deduced from photographs taken at the Radcliffe Observatory, Oxford.

F.R.S., Radcliffe Observer.

Obervers' Remarks:

1909

Nov. 5. Search was made by W. with the 18-inch telescope for Halley's Comet, and it was soon discoverd with power 290 as a faint luminous haze. B. also saw it definitely. Eventually Dr. R. and H. B. also viewed the object, when its motion amongst the stars was evident. An observation for position, from circle readings only, gave the following result: G.M.T. $12^h9^m49^s$: Local Sidereal $3^h3^m4^s$: Apparent R.A. $5^h44^m8^s$: Apparent N.P.D. 73° 9.

Nov. 17. The comet appears dull, with a central nucleus occasionally brightening up. Suspect an elongation, P.A. 135°. The comet is easily seen in the field of the 7-inch telescope with power 90. Conjunction of faint'star with comet observed at $5^h5^m12^s$ Local Sidereal: star very slightly south.

Nov. 18. Nebulous haze, 12″ in diameter. Nucleus precedes centre and is south.

Nov. 22. Conjuction of a 10 mag. star with comet at $3^h42^m10^s$ Local Sidereal. and at about 2′ north.

Nov. 23. Comet about 20″ diameter: condensation diffused. Brigth moonlight.

Dec. 4. Nucleus 10′5 mag. Nebulosity traced for 25″ or slightly more. Short tail suspected, P.A. 135°.

Dec. 18. Nucleus 11 mag.

Dec. 31. Mag 11′5. Diameter of nebulosity about 30″. Comet just visible in the 3-inch finde, power 23.

1910

Jan. 10. Nucleus seen intermittently only. Magnitude 11'5. The photographic image was found to be faint, illdenned, and difficult to measure. The place given is deduced from the mean of measures made by four observers.

Jan. 12. Magnitude of condensation, 12. Conjunction of a 10 mag. star at $2^h18^m6^s$ Local Sidereal, and 45″ north.

Jan. 19. Condensation, magnitude $11^1/_2$. Moonlit sky.

Feb. 2. Magnitude of nucleus $10^1/_2$ or 11, stellar at times Coma 40″ in diameter.

Feb. 3. Mag. 11. Sky slightly hazy.

Feb. 4. Image diffused.

Feb. 7. Mag. 10.

Feb. 11. Mag. 10 or $10^1/_2$. Faint elongation, P.A. 85°.

Radcliffe Observatory, Oxford: 1910 April 13.

Monthly Notices of R.A.A. LXX, 6 (1910) 497.

March, 1910

C 2161, March 1.650, is a rather weak negative. There is strong emission in p.a. 217° (or else the nucleus is elongated) which is approximately the same as the night before. The tail is hardly visible. The outer envelope is much more uniform than on February 28. Its diameter is about 60″. P.a. of the tail is 68°; the southern streamers of the tail are much brighter than the northern. The stubby tail in both sides of the jet from which the slender tail issues has its northern boundary at p.a. 53° and is about 1′ long. Its southern boundary is ill-defined. The nucleus is fuzzy, about 7″ in diameter.

C 2163, March 2.634, shows the jet generating the streamers in p.a. 65° (p.a. of the prolonged radius vector was 59°). The stubby tail is now of the same intensity on both sides of this jet; its edges are in p.a. 26° and 93°. The length of the jet and of the stubby tail is 60″. There is strong emission in p.a. 203°. It is now 20″—25″ long but the central part is much stronger and is some 10″—15″ long. This is now the brightest part of the head. The nucleus is from 3″ to 5″ in diameter.

The strong emission described here is evidently the same as that of February 28 and March 1. Its p.a. changed in four days from 229° to 217° and to 203°.

C 2165, March 5.645, shows a remarkable slightly-curved jet of 35″ length in p.a. 85°. Its tip is bifurcated, and it is only slightly less bright than the nucleus itself. The tail is in p.a. 49° and of about 20° opening. The p.a. of the prolonged radius vector for this date is 57°, so that the tail deviates about 8° from this direction. There is no evidence of the stubby tail or of any other emission. The head is about 1ʹ.5 in diameter. The nucleus is large and fuzzy, about 10″ in diameter.

Visual observation by Quénisset and Baldet,[37], Juvisy, of March 5.265 (?), shows two jets issuing from the head and forming an angle of 70°, one side being much brighter than the other. The comet looks like the letter V. Probably one of the jets was the curved jet of March 5 in p.a. 85°, and the other one the jet of the following date in p.a. 150°. Unfortunately, the p.a. is not given by the observers.

On C 2168, March 7.646, the tail is of about 24° opening; its axis in p.a. 61°. There is a faint streamer 154″ long in p.a. 85°. This is probably what became of the curved jet of March 5. On the assumption of motion in the radius vector this gives the average velocity 2.4 km/sec. There is also strong emission in p.a. 150°. The nucleus is about 5″ and the coma 2′ in diameter.

March 8.614. The nucleus is large and hazy and must be about 10″ in diameter. The coma is visible up to 1ʹ.5 from the nucleus. Short tail.

March 9.646; March 10.642; March 11.643, are badly fogged; they show a large nucleus.

From the middle of March to the middle of April the comet was lost in the Sun's rays. When it became visible again it was of about $2^{m}.5$. (Report)

The heliocentric distance of the comet was 2.4 A.U. on Dezember 6, 2.0 on January 5, 1.5 on February 4, 1.1 on March 6. In spite of these large distances the comet showed remarkable activity in its variation of brightness, formation of emission sectors and jets, and motion of matter around the nucleus in general. The first evidence of a tail was given on February 4, when the comet was 1.5 A.U. from the Sun.

[37] See drawing in Bull. Soc. Astr. France 24, 192, 1910.

N. T. Bobrovnikoff, Publ. of the Lick Observatory, Vol. X VII, pt. 2, p. 338/339, 1931.

April, 1910

April 10. Comète brillante, noyau de 10″; son éclat est comparable à celui d'une étoile de 7e gr.

F. Gonnessiat, Astron. Nachr., Nr. 4453, Bd. 186, S. 205, 1910.

The first observation on Mount Hamilton after the emergence of the comet from the Sun's rays was made on April 12.032. . . . It shows only a trail with a central condensation corresponding probably to the nucleus. It is about 4″ in width. Nebulosity extends about 20″ on one side of the nucleus. A sketch by Mascart[38] at Teneriffe shows some irregular formations around the nucleus.

[38] Bull Soc. Astr. France, 24, 387, 1910.

N. T. Bobrovnikoff, Publ. of the Lick Observatory, Vol. X VII, pt. 2, p. 339, 1931.

April 13. An even round nucleus.

R.T.A. Innes, Astron. Nachr., Nr. 4420, p. 57, 1910.

April 13. . . . Photographed through thin clouds. Tail 33′ long, divided into three branches.

H. E. Wood, Transvaal Observatory Circular, Nr. 4, 11. July, 1910.

April 13.017, . . . There is a bright circular condensation and also a round envelope, each 20″ in diameter. A sketch by Mascart (ibid) shows a very small nucleus with several envelopes.

The first appearance of "parabolic" envelopes is on April 15.014, C 2186. There is only one such envelope on that date. The nucleus is small (2″—3″) and surrounded by an irregular nebulosity extending 20″. The axis of one branch of this nebulosity coincides with the axis of the tail, the other axis is in p.a. about 140°. No measurement of p.a. could be made because of the lack of comparison stars on the negative. The envelope does not appear to be parabolic: vertex 32″, semi-latus rectum 47″ from the nucleus. The phase-angle was about 43°, so that the distortion of the paraboloid might be due to the conditions of projection. The branch

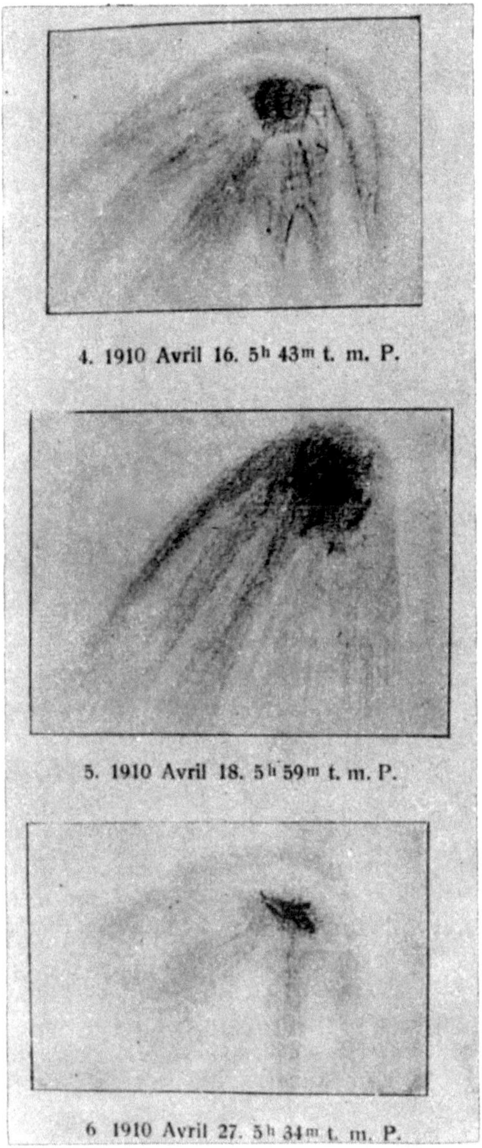

Fig. 1. Head of Halley's comet, photographic observation by I. Mascart, Astron. Nachr. **186** (1910) Nr. 4461, p. 344

of the tail into which emission from the head is directed is the fainter one, contrary to all expectations.

April 16, observations see Fig. 1, No. 4 and Fig. 2, No. 1.

C 2188, April 16.011, shows a bright nucleus 6″ in diameter. There are two

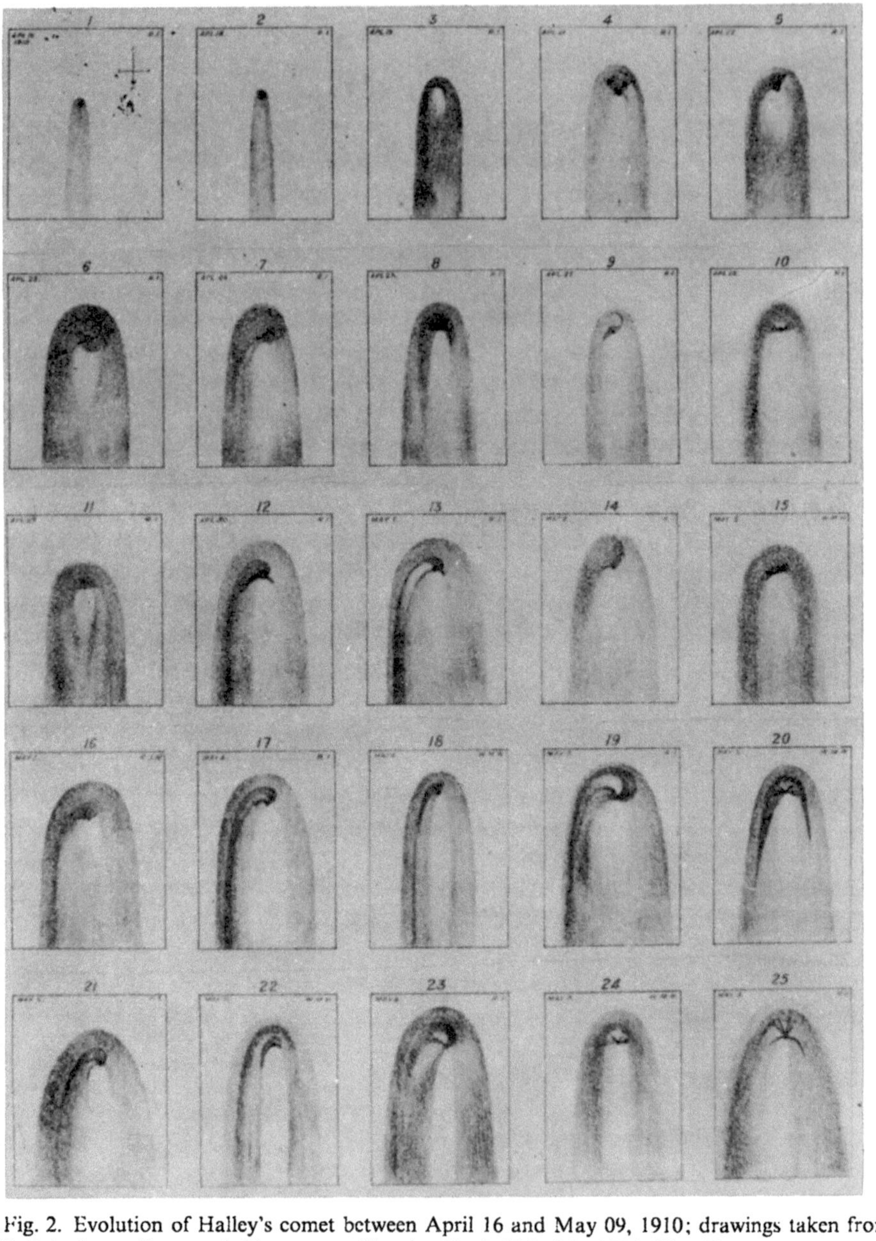

Fig. 2. Evolution of Halley's comet between April 16 and May 09, 1910; drawings taken from R.T.A. Innes, Transvaal Observatory Circular No 4, 11th July 1910, Plate X

emissions. One almost as bright as the nucleus itself is in p.a. 312°, the other is in the shape of a fan, the edges of which are in p.a. 105° and 135°, that is, the axis is in p.a. 120°. There is a bright halo with the nucleus in the cenfer; it is 53″ in diameter. Two envelopes can be seen. The outer one has its vertex 57″ and semi-latus rectum 75″ from the nucleus, that is, more nearly parabolic than the night before. The axis of symmetry of the envelopes is in p.a. 264° (p.a. of the prolonged radius vector is 263°). The northern branch of the outer envelope is sharper and brighter although the main emission goes southeast. The envelopes are not wholly structureless; a suggestion of streamers may be seen.

Innes at Johannesburg, April 16.625, describes a secondary nucleus 12″ from the primary in p.a. 48°3. This must have developed between April 16.0 and April 16.6 as the Crossley plate does not show any unusual condensation in the *nf* quadrant.

Helwan, HR1, April 16.600, shows a very bright halo 84″ in diameter which entirely conceals the details around the nucleus. There is hardly any doubt that this halo is identical with the one on C 2188 described above. If so, the velocity of radial expansion is 0,3 km/sec.

N. T. Bobrovnikoff, Publ. of the Lick Observatory, Vol. X VII, pt. 2, p. 339/340.

On C 2190, April 17.009, several bright emissions are noticeable. There is a jet, very bright, almost as bright as the nucleus itself, in p.a. 313°, nearly the same as the night before, about 20″ long. Another of about the same length is in p.a. 143°. There is a faint suggestion of a secondary nucleus 20″ from the primary in p.a. 20°. It looks like a ball 40″ in diameter; that is, the primary nucleus is on its periphery. This probably the expanded secondary nucleus described by Innes on April 16. The velocity of expansion would be approximately the same as found before for the halo.

The halo on C2190 is broken up. Its diameter ist 80″ and may be considerably larger as it shades off in all directions. The nucleus is very small, about 3″ to 4″ in diameter. There are two streamers making an angle of 30°. The axis of symmetry is in p.a. 264°, the same as on the preceding date. The envelopes are somewhat distorted. There is evidence of motion of matter in the envelopes.

N. T. Bobrovnikoff, Publ. of the Lick Obs., Vol. X VII, pt. 2, p. 340

April 18, observations see Fig. 1, No. 5 and Fig. 2, No. 2

April 18. — Plate I, No. 2. — Tail 1° 45′ long, extending to the edge of the plate used. The tail is composed of five almost straight streamers, apparently radiating from a point just behind the nucleus.

H. E. Wood, Transvaal Obs., Corcular No. 4, 11th July, 1910

April 18. — (a) Plate X, No. 2. — Nucleus elongated 3 × 2 in 113°, 3.′4 by 2.′0 by very rough measures. There is a secondary nucleus at 154° .. 5″ ... The tail is 4° long, and at 3/4° from head is 16′ in diameter. There is an extension of nebulosity towards the Sun. The south side of the tail is the brighter.

R.T.A. Innes, Transvaal Obs. Circular No. 4, 11th July, 1910, p. 28

The Crossley plates of April 18 show a condensation in the northern branch of the inner envelope slowly moving out into the tail. It could be traced on six plates. Table 7 given the observed coordinates α, δ and heliocentric coordinates R, w of condensation and r, v of the nucleus. The times ar corrected for planetary aberration, as in all similar cases in this memoir.

Table 7

1910.0

Point	GMT	α	δ	R a.u.	w	r a.u.		Remarks
1 April	17.002	23h53m47s	+7°49′	0.5926	−8°18′	0.5901	−8°12′	C 2190
2	17.999	53 00	+ 48	.5896	−5 16	.5883	−5 9	C 191
3	18.584	52 23	47	.5901	−3 26	.5877	−3 22	HR 2
4	19.009	52 05	47	.5907	−2 13	.5872	−2 04	C 2193
5	19.582	51 38	45	.5889	−0 6	.5872	+0 18	HR 3
6	19.995	51 12	46	.5926	+0 46	.b872	+0 58	C 2196

The determination of the position of the condensation on C 2196 is hampered by two fine streamers intersecting the condensation. The tail by this time begins to break up into many fine streamers. The analogous condensation in the southern branch was much more diffuse but its motion seems to have been parallel to the northern one.

N. T. Bobrovnikoff, Publ. of the Lick Observatory, Vol. X VII, pt. 2, p. 340, 1931.

The orbit is highly satisfactory and there is no doubt that the effective repulsive force in the envelope was near zero. Therefore $(1 - \mu) = 1$ . . .

The Helwan plate, HR 2, April 18.591, has the brighter inner halo 156″ in diameter. This gives the average velocity of radial expansion 0.7 km/sec, in good agreement with other values. This Helwan plate shows that the emissions from the nucleus which form the inner envelope (studied above) begin to break up into a multitude of fine sharp streamers. . . .

April 19, observation see Fig. 3, No. 3

C 2193, April 19.016, shows a strong emission 39″ long, in p.a. 294°. It is of the same character as the jet of April 17. The position of the jet on C 2193 coincides (+180°) with the direction of elongation of the nucleus given by Innes on April 18.625. The nucleus itself seems to be elongated in that direction, but measurement is difficult owing to overexposure. There are two indefinite globular condensations, 20″ to 25″ in diameter, tangent to the nucleus. Their centers are in p.a. 51° and 179°. The latter is probably Innes' expanded secondary nucleus. This secondary nucleus is better visible on C 2194, April 19.022; its center is in p.a. 179°, about 12″ from the primary. Assuming that this secondary nucleus, a point on April 18.625, became a globe of 12″ radius on April 19.022, we obtain for the average velocity of expansion 0.3 km/sec.

The jet in p.a. 294° can be seen on Innes' drawing.

N. T. Bobrovnikoff, Publ. of the Lick Obs., Vol. X VII, pt. 2, p. 341/342; 1931.

C 2197, April 20.018, shows a strong concentration of material in p.a. 57°. It may be described as a globe of 40″ diameter. The primary nucleus is on its periphery. The bright jet of the preveding night has considerably diminished in brightness but gained in length. There is a faint halo 3.′0 in diameter with the nucleus in the center.

Plate C 2196, April 20.002, is . . . the last plate on which condensation E in the northern branch of the envelope is still distinguishable.

Objective prism spectrogram W 199, April 20.007, has diffused images among which can be recognized CN IV and C IV.

A remarkable phenomenon developed in the tail on April 21. The northern branch of the tail, the same in which condensation E was moving, became twisted and knotted as if a sudden explosion had taken place there. This is best seen on the Kodaikanal photograph, [40] . . ., and the Santiago plate. In spite of the short interval of time (only $0\overset{d}{.}427$) between these two exposutes, the details of the Kodaikanal plate are hardly recognizable on the Santiago plate. The two photographs have been reduced to the same scale to show the motion in the tail better. There were found altogether six negatives and photographs in which the details in the tail could be measured. Condensations A, B, C, D, are shown in Plate 10, figure 8. Table 10 gives the result of measurement.

Table 10

Point	GMT April	A 1910.0 α	A δ	B 1910.0 α	B δ	C 1910.0 α	C δ	D 1910.0 α	D δ	Remarks
1	20.998	$23^h50^m02^s$	+7°60′							
1	20.998	$23^h50^m02^s$	+7°60′							WS 12
2	21.472	48 44	49	$23^h48^m38^s$	+7°52′	$23^h47^m11^s$	+7°55′	$23^h47^m59^s$	+7°42′	K
3	21.621	47 55	46	48 01	7 52	45 58	51	47 33	43	J 4
4	21.900	46 22	45	46 32	7 51	43 25	48	46 11	38	S 1
5	21.995	45 45	45							WF 16
6	22.621	39 12	26							J 5

The heliocentric coordinates of the condensations and of the nucleus are given in Table 11.

Table 11

Point	r	v	A R	A w	B R	B w	C R	C w	D	D R	D w
1	0.5878	+3°45′	0.6953	+2°27′							
2	.5885	5 29	.6144	3 28	0.6198	2°42′	0.6395	+1°10′	2	0.6118	4°38′
3	.5888	5 57	.6174	4 10	.6253	2 49	.6447	1 51	3	.6168	4 38
4	.5893	6 49	.6297	4 21	.6371	3 04	.6651	1 54	4	.6191	5 54
5	.5894	7 06	.6357	4 15							
6	.5907	9 00	.6703	6 27							

[40] Obtained by Evershed and first reproduced in Monthly Notices, 70, pl. 20, 1910.

N. T. Bobrovnikoff, Publ. of the Lick Obs., Vol. X VII, pt. 2, p 342/343, 1931.

CONDENSATION A ... $(1 - \mu) = 54.4 \pm 4.1$ ...

CONDENSATION B ... $(1 - \mu) = 31 \pm 30$. Probably the repulsive force in this condensation was the same as that in condensation A, to which it was adjacent. ...

CONDENSATION C ... $(1 - \mu) = 238 \pm 54$. ...

There is no doubt that condensation C was moving under a much greater repulsive force than either A or B, which might be surmised from its very rapid motion. ...

Condensation D showed hardly any repulsive force, which might be due to the difficulties of measurement. If there was any repulsive force, it was very small. ...

It seems that condensation A underwent some internal transformation shortly before the first moment of observation, that is, April 20.998. This transformation may be described as an explosion in the tail. The explosion changed the motion of the condensation to such an extent as to remove its perihelion from the nucleus. This is still more evident in condensation C, the perihelion of which must have occurred at approximately 0.05 A.U. from the nucleus.

If we consider now the motion in envelope E, we shall find that the ... closest approach of A and E occurred about April 20.5. At this moment E was almost halfway between A and the nucleus. It seems permissible, therefore, to think that A and other condensations originated at about April 20.5 from envelope E. A sudden transformation, resembling an explosion, took place whereby the small repulsive force in the envelope (of the order of 1) was changed into various repulsive forces from 30 to 240 times gravitation.

N. T. Bobrovnikoff, Publ. of the Lick Obs., Vol. X VII, pt. 2, pp. 343—346.

April 20. — Not reproduced. This morning was very cloudy, but an exposure of about 2 minutes was secured as a cloud rift passed across the place of the Comet. A very clear image of the nucleus was obtained, and there seemed to be a greater outpouring of matter into the tail on the North edge.

April 21, observation see Fig. 2, No. 4

April 21. — Plate I, No. 4. — Time of exposure, 14 h. 42 m. to 15 h. 27 m. G.T. This photograph shows a very extraordinary tail, the irregularities in being very much greater than those shown on any other photograph. The irregularities are chiefly on the North edge of the tail; a bright streamer forming the North edge of the tail bends sharply twice at right angles about 45′ from the head, and further along are several smaller contortions. The main portion of the tail consits of four narrow streamers, forming two pairs; the Northern pair tail about wavy. A small brush leaves the South edge of the tail about 30′ from the head. The internal structure of the tail is very confused and contorted at 45′ from the head.

April 21. — (a) Plate X, No. 4. — Secondary nucleus seen, but indistinctly. The two nuclei were involved in the haze of the coma. A dark vacancy immediately behind the main nucleus.

(b) Secondary nucleus or slight condensation at 123° ±.

The general aspect of the tail certainly favours the idea of an explosion ... The north edge of the tail, smooth on April 21,480, became jagged and contorted on April 21.907 with a great number of secondary centers of condensation. It appears as if matter from A was expelled in all directions. The most interesting circumstance is that while streamer A increased in diameter during the interval not less than ten times (and consequently in volume at least 100 times), its brightness (in comparison with the head) did not diminish but, on the contrary, considerably increased. It is difficult of course to compare a reproduction with an original, but we shall see later that this is a common phenomenon in such explosions. Gaseaus matter becoming rarefied becomes thereby more luminous, which suggests some chemical process akin to explosion ...

Transvaal Obs., Circular No. 4, 11th July 1910, p. 24, 28.

April 22, observation see Fig. 2, No. 5

C 2198, April 22.002, shows an elongation of the nucleus or a strong emission in p.a. 313°. There is a short emission or a secondary nucleus in p.a. 157°, 10″ long. The nucleus itself is illdefined, 10″ ± in diameter. There are two envelopes. The inner one, brighter, has dimensions 35″ × 47″ (vertex × semi-latus rectum); the outer one is 62″ × 113″. The ratio of the semi-latus rectum to vertex in the first case is 1.34, in the second, 1.82. If this distortion of the paraboloid of the head were due to projection, both the ratios should be the same.

N. T. Bobrovnikoff, Publ. of the Lick Observ., Vol. XVII, pt. 2, pp. 346, 347, 1931.

Secondary nucleus 160° and close to primary, like a big ill-defined double star. South branch of tail mach the brighter.

R. T. A. Innes, Astron. Nachr. Nr. 4411, S. 57, 1910.

C 2199 of April 22.017 shows a strong halo 58″ in diameter. Apparently the same halo on Helwan HR 5, April 22.586, has a diameter 60″. There seems to be no appreciable expansion.

C 2200, April 22.999, is reproduced in Plate 10, figure 7. This plate and C 2201, April 23.015, show a bright nucleus 4″ in diameter surrounded by a spherical bright haze about 10″ in diameter. There is a short bright jet again in p.a. 313°. This jet is well shown on Helwan, April 23.584 (HR 6), in the same p.a. The inner envelope on the Crossley plates is 38″ × 62″, the outer 89″ × 129″; that is, the ratios of semi-latus rectum to vertex are 1.63 and 1.45.

The interesting feature of the first week after perihelion was the persistent jet in p.a. 313°, first notical on April 16.011 and last on April 23.584. It was very bright, but of varying intensity. The p.a. of the prolonged radius vector was 263°19′ on April 16.0, 260°42′ on April 20.0, and 258°46′ on April 24.0; that is, it was changing very slowly. Thus, the jet kept approximately the same position in reference to the radius vector. This is certainly not in accordance with the idea of a rotation of the nucleus. Small, irregular oscillations might have occurred; on April 19.016 the p.a. of the jet was measured 294°, which is decidedly too small to be explained by errors of measurement. This jet apparently supplied the material for the northern branch of the tail. Especially, condensation E of April 17—20 may have originated from it.[41]

[41] This jet was noticed by many obervers. See for instance Philips (Jour. Brit. Astr. Assoc. 20, 426, 1910) and Eddie (ibid. p. 427). When it ceased, the southern part of the tail became the brighter.

N. T. Bobrovnikoff, Publ. of the Lick Obs., Vol. X VII, pt. 2, p. 347, 1931.

April 23, observation see Fig. 2 No. 6

April 23. — (a) Plate X, No. 6. — Nucleus much smaller, the outpouring is all on the S. side; no secondary nucleus seen.

(b) Nucleus fairly well defined, streamers on S.f. side. S. tail much the brighter.

Transvaal Obs., Circular No. 4, 11th July, 1910.

No secondary nucleus visible. The outpouring is concentrated on the south side.

R. T. A. Innes, Astron. Nachr., No. 4420, p. 57, 1910.

April 24, observation see Fig. 2 No. 7

... The nucleus is very sharp and small, about 4″ in diameter, and is surrounded by a bright nebulosity 10″ in diameter. Several faint jets may be suspected. The general appearance is that of the preceding date but the south side of the envelope is now much the brighter.

N. T. Bobrovnikoff, Publ. of the Lick Observ., Vol. XVII, pt. 2, p. 347, 1931.

Secondary nucleus 80°, 7″. Both nuclei more stellar.

R. T. A. Innes, Astron. Nachr., Nr. 4420, p. 57, 1910.

On distingue dans le noyau deux jets lumineux de 5″ et 8″ de longueur. —

F. Gonnessiat, Astron. Nachr., No. 4453, S. 203, 1911.

April 25, observation see Fig. 2 No. 8

April 25. — (a) Plate X, No. 8. — The comet was much brighter this morning. No secondary nucleus seen.

(b) Comet much brighter. Nucleus larger and fairly well-defined. Tail more evenly developed, but S. tail is still the brighter and longer.

Transvaal Observ., Circular No. 4, 11th July 1910, p. 28.

"Strong somewhat rounded protuberance toward the Sun, seen also visually" (Curtis). The outpouring on this date was mostly to the south but there were no distinct jets.

N. T. Bobrovnikoff, Publ. of the Lick Observ. Vol. X VII, pt. 2, p. 347, 1931.

Despite the full moon the comet's head and a part of its tail were visible to the naked eye at 15^h30^m. The jet observed on April 24 had vanished, but the south side of the tail, near the head, appeared more extensive than the north side. It was visible to a greater distance from the head (going lengthwise down the tail), and was broader at the head. The nucleus was therefore excentric in the head, being the nearer to the north edge of the head. At a point in this north edge of the head, near the nucleus, the outline of the coma seemed to bend inward a little, just as if an invisible body of circular form were coming up to occult the comet's head. The finder corroborated the 20 inch as to the brightness and greater extension of the south side of the tail. —

H. A. Howe, Astron. Nachr., Nr. 4784, S. 134/135, 1912.

April 26

Der Komet wurde in Potsdam am Steinheilschen Refraktor von 13.5 cm Öffnung und 216 cm Brennweite mit nicht ganz 80-facher Vergrößerung seit April 26 an allen Morgen, an denen es die Witterung überhaupt gestattete, von seinem Aufgange an beobachtet, bis er in der Dämmerung verschwand, also gewöhnlich von etwa $14^3/_4{}^h$ bis etwas nach 16^h M. Z. Potsdam. Da der Osthimmel fast immer von dichtem Dunst oder Wolken mehr oder weniger bedeckt war, konnte der Komet häufig nur in Wolkenlücken gesichtet werden, und gewöhnlich war die Dämmerung, wenn er sich über die Dunstschicht erhob, schon zu weit vorgeschritten, als daß noch Beobachtungen des Schweifes gemacht werden konnten. Aus gleichem Grunde gelang es nur an zwei Morgen, Mai 1 und 7, für photometrische Zwecke brauchbare photographische Aufnahmen des Kometen zu erhalten.

Im allgemeinen war nur der Kern des Kometen in rötlich gelbem Lichte mehr oder weniger verwaschen mit der Koma sichtbar, so April 26, 29, 30, Mai 1.

W. Münch, Astron. Nachr., Nr. 4425, S. 137, 1910.

At 15^h30^m the comet with its tail was dimly visible to the naked eye; an hour later is was at the limit of naked eye visibility, being fainter than β Pegasi, and of about 3^m. In the telescope the tail looked milky and much brighter than heretofore. No jet was seen. The head looked very symmetrical, except that the northern part seemed not quite so wide and perhaps fainter that the southern side. At 16^h30^m the twilight had about blotted the northern side out, while the southern was still visible. At 16^h46^m nothing could be seen but the nucleus and a very short extension of the coma south of the nucleus. In trying to test the relative brightness of the northern and southern sides of the head, the observer's head was turned in various directions, but the southern side of the comet's head always looked brighter.

H. A. Howe, Astron. Nachr., Nr. 4784, S. 134, 1912.

The Santiago Plate S4, April 26.908, can be compared with the objective prism spectrograms OP1 and OP2 (April 26.897 and 26.923). These spectrograms show an extremely fine nuclear line, the average width of which is 0.'3. The CN IV_1 band has a very bright core of 0.'7 diameter and a round nebulosity of 2.'5. CN IV_2 shows the same features, only much gainter. The CN IV bands extend into the tail for about 8′ and then stop abruptly. C IV has only the core of 0.'7. Direct photograph S4 is badly fogged. It shows a solarized central part of the head some 2.'2 diameter, which corresponds to the central nebulosity in the CN bands. There is a bright extension of 8′ length into the tail, which is in reality the edge of the envelope. Apparently it consisted of CN. The tail images of CO^+ can just be made out. They are of rather uniform intensity, although on the direct photograph the edges of the envelope spoken of are very much brighter than the rest of the tail. The envelope must have consisted entirely of CN up to 8′ from the nucleus. The spectrum of the nucleus is of the solar type.

The only other band visible is CN III; it has the same structure as CN IV_1, but is much fainter than even CN IV_2.

N. T. Bobrovnikoff, Publ. of the Lick Observ., Vol. X VII, pt. 2, p. 248, 1931.

Am 26. April wurde der Komet auf unserer Beobachtungsstation am Pedrogilpaß (Höhe nahezu 2000 Meter) zum erstenmale mit bloßem Auge gesehen. Trotz hellen Mondscheins war ein deutlicher Schweif sichtbar. Kern ganz sternartig, mit bloßem Auge etwa 3. Größe geschätzt. — Auf derselben Station wurde am 27. April eine photometrische Messung der Kernhelligkeit mit einem Zöllnerschen Astrophotometer ausgeführt, welche (ohne Berücksichtigung der Extinktion) ergab, daß der Kern etwa um 1 Größenklasse heller war als der Stern d Piscium. Am 30. April, 1. und 3. Mai wurden zahlreiche Aufnahmen des Kometen mit der Stereoskop-Kamera bei verschiedenen Expositionszeiten gemacht. Die Helligkeit des Kerns nahm von Tag zu Tag zu. Sie wurde am 3. Mai um etwa $0^m.1$ bis $0^m.2$ höher geschätzt als α Pegasi. Der Schweif war trotz des hellen Mondscheins zuletzt über die Verbindungslinie „Venus — α Pegasi" hinaus zu verfolgen.

Auf unserer höheren Beobachtungsstation „Alta Vista" (3300 Meter) war die Witterung so günstig, daß vom 7. bis zum 14. Mai, als wir die Beobachtungen abbrachen, ohne Ausnahme an jedem Tag Aufnahmen des Kometen teils mit der Stereoskop-Kamera, teil mit einem Görzschen Anastigmaten von 180 mm Brennweite gemacht werden konnten. Während dieser ganzen Zeit war der Komet in der reinen und durchsichtigen Höhenluft schon bald nach seinem Aufgang eine prachtvolle Erscheinung. Die Helligkeit des Kerns wurde an den letzten Tagen weit höher als die eines Sterns 1. Größe geschätzt, die Länge des Schweifs erreichte am 7. Mai nach den Schätzungen mit bloßem Auge mindestens 20°, am 10. Mai mindestens 30° und am 14. Mai bei etwas dunstiger Luft etwa 50° bis 60°. In der Nacht vom 18. zum 19. Mai, die wir in Orotava zubrachten, war der Himmel mit leichtem Gewölk bedeckt; irgend eine auffällige Erscheinung konnte nicht wahrgenommen werden.

Die Teneriffa-Expedition hat von neuem einen Beitrag zu der Lehre geliefert, daß man bei ähnlichen Erscheinungen wie der Halleysche Komet keine Mühen und Kosten scheuen sollte, um günstige Beobachtungsstationen insbesondere in größeren Höhen über dem Meere aufzusuchen und die betreffenden Expeditionen mit den allerbesten instrumentellen Hilfsmitteln auszurüsten.

G. Müller, Astron. Nachr., Nr. 4425, S. 136, 1910.

April 27, observations see Fig. 1 No. 6 and Fig. 2 No. 9

La tête de la comète est principalement formée d'un noyau lumineux, de forme elliptique, dont le grand axe, dans le plan N—S, se prolonge symétriquement en s'incurvant vers l'ouest sur 2′ à 3′ d'étendue. Absence de queue. —

P. Chofardet, Astron. Nachr. Nr. 4489, p. 4, 1911.

Während meiner Reise in Süd-Rußland gelang es mit viermal den Halleyschen Kometen zu beobachten:

1910 April 27, Nicolajew, $15^1/_2{}^h$ Petersb. Zeit. Kern 3. Größe, Schweif kaum sichtbar.

K. Pokrowsky, Astron. Nachr., Nr. 4420, S. 59, 1910.

Der Halleysche Komet konnte in Königstuhl zum erstenmale als Morgenstern heute früh bei hellem Mondschein beobachtet und photographiert werden. Herr Lorenz verglich die Helligkeit des Kerns mit ω Piscium und α Pegasi und fand ihn etwa 3. Größe. Herr Lorenz sah am 6 Zöller (Vergrößerung 36fach) einen deutlichen Ausströmkegel und einen etwa im Parallel gerichteten kurzen Schweif, der auch im Sucher sichtbar war. Der Komet war trotz Mondschein und Dämmerung leicht mit bloßem Auge zu sehen und konnte noch 4^h30^m im Operngias gefunden werden. Es wurde eine Aufnahme des Kometen von Herrn Lorenz gemacht, während Herr Ernst einen Anschluß an einen Stern 7. Größe erhielt.

M. Wolf, Astron. Nachr., Nr. 4411, S. 319, 1910.

April 28, observation see Fig. 2, No. 10

Plate X, No. 10. — Nucleus not quite round. Butterfly appearance. S. branch of tail somewhat brighter than N.

Transvaal Obers., Circular No. 4, 11th July, 1910, p. 28.

C 2207, April 28.994, shows on the south side a sharp short streamer; from its tip, A, issue several fine streamers especially well visible on Helwan, April 28.590 (HR9). The motion of this tip can be followed on three which makes it possible to determine $(1 - \mu) \ldots = 23 \pm 0.3$. This value is of course not very accurate, in spite of its small probable error. The interval of time covered by the observations is too short ... Several streamers are seen. ...

April 29, observation see Fig. 2 No. 11

C 2208, April 29.010, can be compared with the objective prism spectrogram OP3, April 28.908. The spectrum looks very much like that on April 26. There is the same fine nuclear line of 0'.3 width. The diameter of the core of CN IV is 0'.6. The outer halo around this band is 1'.7 in diameter. CN IV extends into the tail 10'. The Crossley plate shows a solarized nucleus of 0'.8 diameter corresponding to the core of CN IV. A very bright circular halo of 1'.8 evidently corresponds to that of CN IV. The bright edge of the envelope generating streamers A extends up to 10' into the tail; that is, it consists mostly of cyanogen. The Swan C IV band is restricted to the nucleus. The tail images of CO^+ are of uniform brightness, whereas on the direct photographs the envelope generating streamers A is much brighter than the rest of the tail (see Pl. 10, fig. 10).

The formation of streamers A is a good illustration of how cyanogen in the head is transformed into CO^+ of the tail. We shall see that this was a common occurrence with Halley's Comet. The exact meaning of such "transformation" will be discussed later.

V. M. Slipher has noticed the same behavior of CN and CO^+ during this period. On April 25 he notes: "While knot 3883 spreads slightly, it is not a constituent of the tail, nor are any of the other prominent head emissions; and conversely, the tail streamers are not obviously represented in the spectrum of the head." On April 19: "A streamer extends from the less refrangible and brighter edge of this knot (λ 3883) faintly into the tail for two-thirds of a degree." April 30: "The streamer λ 4010 (CO^+) ... can be seen at a distance 10° from the head. ... The double head emission at λ 3883 is diffused for half a degree into the tail"'

N. T. Bobrovnikoff, Publ. of the Lick Observ., Vol. X VII, pt. 2, p. 349.

(a) Plate X, No. 11. — Nucleus more symmetrical. S. branch of tail somewhat brighter than N. branch.

(b) Nucleus small and well-defined. Slightly elongated about 160°—340° with "fans" of stramers at each end. S. tail slightly brighter than N.

Transvaal Observ., Circular No. 4, 11th July, 1910, p. 28.

Observations of Halley's Comet made on board s. s. "Miami"

By *W. G. Preece*, Second Officer (Communicated by Capt. *Campbell Hepworth*).

On April 29, at about 7 O'clock a.m. G.M. time, in latitude 26° 56' N., long. 52° 50' W., I observed a very bright comet, rising in close proximity to Venus and λ Pegasi. From that date onwards it became brighter and the tail longer each day, until in May 16 its tail reached from the horizon to the zenith. It seemed to be travelling towards the north and west. After May 16 we were unable to see either the comet itself or the tail on account of daylight breaking so early.

Meteorological Office, London, S.W.: 1910 June 6.

Monthly Notices of R.A.S. LXX, 8 (1910) 621.

April 30, observation see Fig. 2, No. 12

Plate X, No. 12. — An ill-defined secondary nucleus or bright extension in 120°.

Tail 30′ wide at 5° from nucleus. Tail consists chiefly of three bright parallel rays. Leaving the North edge of tail near the head is a series of short parallel streamers about 1° long.

Transvaal Observ., Circular No. 4, 11th July, 1910, p. 25, 28.

May, 1910

Halley's Comet, 1910: Observations made in Jamaica

By *Maxwell Hall*

From February 1 onwards the circular head and nucleus were seen through a small telescope at Helfway Tree near Kingston, but observations were not commenced until the disappearance of the Moon from the morning sky early in May.

May 6, 4 a.m. Chapelton. The Moon, three days before new, was rising; a small sketch was made showing the tail to be about 12° in length, and to make an angle of about 30° with the vertical to the right. Dawn occurred at 4.30 a.m.

May 8, 4 a.m. Chapelton. Another sketch was made showing the tail to be about 18° in length, and to make the same angle with the vertical as before.

May 14, 4 a.m. Halfway Tree. The tail extended to θ Pegasi, the central line being 1° south of that star, so that the end of the tail was at R.A. $22^h 6^m$ Decl. +5°, and its length was 46°. A sketch was made of the nucleus through the 3-inch telescope, showing it to be of a triangular form.

May 16, 4 to 4.30 a.m. Halfway Tree. The tail extended to the small group of stars 68, 69, 70, and 71 Aquila; it was a little south of the group. The tail was straight, about 3° broad, and of a silvery white colour; the head was brighter than a star of the 1st magnitude, but far below the brightness of Venus. The central line of the tail passed about 1° south of θ Pegasi as on the 14th.

It was deduced that the end of the tail was at R.A. 20^h30^m Decl. $-4°$, and that its length was $80^1/_2°$.

May 17, 3 a.m. Chapelton. The tail extended to the same group nearly, but the middle of the tail was at the middle of the group. The upper edge of the tail pointed to θ Aquila. At θ Pegasi its width was about half the distance between that star and α Aquarii, or $3^1/_2°$.

It was deduced that the end of the tail was at R.A. 20^h35^m Decl. $-2^1/_2°$, and that its length was 87°.

The head could not be seen from the place of observation, but it was seen at another place in Chapelton from 4.20 a.m., when it rose, until 4.35 a.m., when dawn overpowered it.

May 18, 4 to 4.20 a.m. Chapelton, at the hospital grounds for better view. The central line of the tail passed $^1/_2°$ south of γ Pegasi and over θ Pegasi. The northern edge touched θ Aquila, and it reached nearly to the Milky Way. Its breadth at γ Pegasi was about 5°, and at θ Pegasi 4°.

It was deduced that the end of the tail was at R.A. 19^h30^m Decl. $-4°$, and that its length was 113°.

Dawn occurred at 4.25 a.m. As the comet's head would not rise until 4.53 a.m., and as daylight interfered at 5.5 a.m., it was not thought remarkable that the head was not seen, especially als there were some light clouds over the eastern horizon.

May 19, 3 to 3.20 a.m. Chepelton. As the tail was supposed to have swung round to the western sky during the night, it was surprising to see the tail still in the eastern sky. It was broader and dimmer than before, and somewhat resembled the Zodiacal Light when well seen.

The following obervations were made: — The centre of the band passed over ζ Pegasi and θ Aquila, and its end was lost in the Milky Way. The centre of the band passed between ε and θ Pegasi, the edge nearly touching ε, and enclosing θ, so that the centre was a little south, and so that its breadth between these stars was 6°.

After 3.20 a.m. it became cloudy, and it was a cloudy evening.

May 20, 3.30 to 4.30 a.m. Chapelton, at the hospital grounds. The southern edge of the band was clear of Pegasi; the northern edge was very ill-defined and lay about 1° beyond α Andromeda. The southern edge touched ζ Pegasi. The band was central between ε Pegasi and α Equulei; also between θ and η Aquilae, its end was lost in the Milke Way.

Between ε Pegasi and α Equulei its breadth was $^3/_4$ the distance between these stars, or 7°; but between γ Pegasi and α Andromedae it was as much as 15°.

The southern edge was very well defined, and there was a dark space between it and the Zodiacal Light, which was seen faintly to the right.

May 20, 7.30 p.m. Montego Bay. A sketch was made of the comet in the west and its new fan-shaped tail, which stretched away from the Sun. It was about 13° in length, and its greatest transverse with was 6°. The head was brighter than γ Geminorum, and of about the 2nd. magnitude.

There was no appearance of any tail below, or of any connection with its former tail. There was moonlight, and there were light clouds to the west; but the observations were reliable as fas as they went.

May 21, 3.40 a.m. Montego Bay. Moon set. Band very faint. It southern edge was 2° norht of γ Pegasi, and toched α and ε Pegasi. Its northern edge was at a Andromeda. Its width at α Pegasi was from that star to I° south of μ Pegasi or 9°.

The Zodiacal Light was pretty distinct, and there was a dark band in the sky between the two.

4 a.m. Very clear. The cometary band was very strong on the N.E. horizon and very broad; it followed the line of the stars α, β, γ Andromeda, and its width was from γ Andromeda to α Arietis, or 19°.

4.10. a.m. There was a strong light to the N.N.E. under Cassiopeia.

4.14 a.m. The band over the N.E. horizon looked exactly like a fine display of the Zodiacal Light; it was of a pearly white colour.

4.22 a.m. Dawn, and the whole band rapidly disappeared.

May 21, 7.10 to 7.30 p.m. Kempshot Observatory. The tail had increased to $28^1/_2$°, and had lost most of its fan-shaped appearance. Its extremity was at R.A. 8^h30^m Decl. + 12°. The nucleus was as bright as γ Geminorum. The head was viewed under different powers, but there was no connection whatever with the former tail.

May 22, 4.18 a.m. Kemshot Observatory. The Moon had just set, but dawn had commenced. A glow was seen above the N.E. horizon, but it was mixed up with the Zodiacal Light on the right and the Milk Way on the left, so that its boundaries could not be made out.

Moonlight prevented any further morning observations of this fading cometary band.

7.30 p.m. The fan-shaped appearance of the tail had all gone; the observed length was about 34°, and this was the maximum for the new tail seen in the west. With a power of 120, the nucleus was only 4‴4 in diameter; and as log Δ was 9′2654, this gave a diameter of only 365 miles.

The subsequent observations made were of no particular interest; on the 30th. the length of the tail was 30°, and then it rapidly diminished.

Deductions were made from these observations, and notices were sent to the local press from time to time; but a more complete study than was then possible is now required.

Monthly Notices of R.A.S. L XX, 8 (1910) 615.

Jamaica: 1910 June 4.

Mai 01

C 2210, May 1.003, shows evidence of rapid motion in the fine streamers in spite of the short interval of time. The streamers are remarkably developed on this night. It was not possible to measure the motion of matter in the streamers, as they changed very rapidly. Thus two streamers widely separated on C 2209 are almost in contact on C 2210. The tail is breaking up into the finest streamers.

The dark lane which forms the axis of symmetry of the tail is in p.a. 258°; that is, it coincides with the prolonged radius vector (p.a. 256°) within the error of measurement. It is not absolutely void of matter, but appears dark chiefly by contrast ...

Crossley plates 2211—2213 of May 1 show a great number of the finest streamers. They change their position and shape very rapidly, as may be ascertained by examining C 2211 (May 1.974) and C 2212 (May 1.996). However, measurement is impossible, due to the wealth of confusing details. As has been observed before, the streamers issue not from the nucleus itself, but rather from some quiescent formation near the nucleus ...

The nucleus is 4″ in diameter (C 2213); the head has (C 2212) several concentric halos of diameter 2.′6, 3.′7, 6.′0, diminishing in brightness with recession from the nucleus.

N. T. Bobrovnikoff, Publ. of the Lick Observ., Vol. X VII, pt. 2, p. 349, 1931.

May 1. — (a) Plate X, No. 13. — S. branch of tail is the brighter. A rift has appeared S. of the nucleus.

(b) Nucleus rather smaller and well-defined, secondary nucleus suspected at about 30°. N. side of head brighter than S. but S. tail brighter than N. On going to the 9-inch, it was seen that the nucleus was elongated slightly in direction of 150°—330°, suspected secondary nucleus is apparently specially strong outpouring of matter. Lane in S. branch of tail, bordered on N. by a streamer.

Transvaal Obers., Circular No. 4, 11th July, 1910, p. 28.

Mai 1, Nicolajew. Der Komet trat aus den Wolken um 15^h20^m. Scharf begrenzter Kern (für das unbewaffnete Auge); Helligkeit $2.^m3$, Länge des Schweifes $1^1/_2$°. Im Fernrohr auf der der Sonne zugewandten Seite des Kerns ein halbkreisförmiger diffuser Schein sichtbar, Kondensation in dem nördlichen Teile desselben.

K. Pokrowsky, Astron. Nachr., Nr. 4420, p. 59, 1910.

May 02; observation see Fig. 2, No. 14

Mai 02, Pos.-Winkel des Schweifs 250°. Kern hat längere Ausstrahlung nach Norden im Positionswinkel 34°, kleinere nach Süden im Positionswinkel 130°. Abstand der Ränder beider Ausstrahlungen 1″. Durchmesser des Kerns 2″. Länge des Schweifs 12°, reicht bis $23°10^m + 6°$. Schweif etwas nach oben gekrümmt. Venus bleibt weit über dem Schweif.

Astron. Nachrichten, Nr. 4501, S. 225/226, 1911

May 2. — Plate X, No. 14. — Nucleus small and stellar, but somewhat indistinct f. No jets, etc., clearly seen. S. branch the brighter.

Transvaal Observ., Circular No. 4, 11th July, 1910, p. 28.

The Santiago plate S6, May 2.894, can be compared with the objective prism OP5, May 2.894. CN IV_1 has a faint envelope extending toward the Sun for 3.′4; away from the Suns extends for a least 5′, graduall fading out. On Helwan, HR 12, Ma 2.583, this corresponds to the envelope and jets which generate fine streamers. It must be noted that C IV, while brighter in the nucleus, does not show this envelope. The inner core of CN IV_1 is very bright; it is about 4′ in diameter and corresponds to the brightest part of the head; the sunward side of this core is brighter, as seen on the direct photographs. CN IV_2 and CN IV_3 have the same structure as C IV_1, but are much fainter. The ratio of their intensities is 8:5:4. The brightest knot of all is C IV; it is 10 on the same scale. The diameters of the subheads in C IV are, however, smaller than the inner core of CN IV, measuring only 2′, the same size as C V and CN III. Knots CN III_1 and CN III_2 have faint jets, directed toward the Sun, about 4′ in length. C V has one, although it is brighter than CN III (5:3). There are no jets even at C IV, which are the brightest knots in the whole spectrum. The nuclear line is from 1.′0 to 1.′5 in width, probaly widened by insufficient guiding.

The CO^+ images are well visible. They show some of the fine details of the streamer tail in the direct photographs. The glow in the CO^+ images extends about 12′ from the nucleus, as on the direct photographs. The brightest part of this glow coincides with the glow at CN IV. It appears as if the CO^+ images begin not at the nucleus itself, but some distance from it, approximately where the bright core of CN IV ends. The space between the nucleus and the end of the CN IV bright core is very faint in the CO^+ streamers.

As on previous occasions the CO^+ images are of uniform brightness, only slightly brighter near the nucleus. However, on the direct photographs the part near the head is enormously brighter than the rest of the tail. This is of course due to the presence of CN IV, which knot has this very bright extension into the tail.

N. T. Bobrovnikoff, Publ. of the Lick Observ., Vol. X VII, pt. 2, p. 349/350, 1931.

May 03; observations see Fig. 2, No. 15 and 16

May 3. — (b) Plate X, No. 15. — Nucleus small, bright, rather ill-defined. Dark space behind head sharply defined on f. side by fan-shaped streamers. S. tail brighter and broader than N. tail. (9-inch). [N.B. — R.I. was absent attending a congress; Mr. R. J. Mitchell was good enough to come up and make a drawing, Plate X, No. 16. Mr. Mitchell adds: "Noticeable features (i) Fanshaped dark space behind the nucleus; (ii) Slightly accentuated stream of superior brightness along S. branch of tail; (iii) Fan inclined to be brighter on S. side."]

May 3. — Plate II, No. 16. — Exposure from 14 h. 37 m, to 15 h. 07 m. Tail 57′ wide at 5° from nucleus. Structure of tail very regular near head, with strikingly symmetrical sets of rays on either side of the axis. The tail is almost divided along the axis, but the division is not distinct. A streamer leaves the South edge of the tail.

Transvaal Observ., Circular No. 4, 11th July 1910, p. 25, 28.

A l ceil nu la comete mesure 20° de longueur d'arc; l'éclat du noyau est 4.5 gr., dans la lunette, on le voit de 6ᵉ gr. On apercoit une étoile de 10.5 gr. à travers la queue; le noyau paraît double ou allongé suivant l'angle de position 108° et coupé par une partie sombre. Le crépuscule favorise l'étude des détails du noyau.

F. Gonnessiat, Astr. Nachr., Nr. 4453, S. 205, 1911.

Kopfhelligkeit im Sucher $2^{m}.5$ bis $3^{m}.5$, sehr verwaschener Kern im Hauptrohr $5^{m}.5$ bis $6^{m}.0$ nach Schätzung gegen 31 Piscium (PD $6^{m}.58$). Schöne Kernausströmungen nach der Sonne zu, die auf der Rückseite einen dunklen Hohlraum einschließen. $16^{h}11^{m}$. Z. Bergedorf, etwa 12^{m} vor Sonnenaufgang, war der Komet trotz feiner Cirrocumuli im Sucher noch gut erkennbar. —

K. Graff, Astron. Nachr., Nr. 4539, S. 45/46, 1912.

May 04; observations see Fig. 2, No. 17 and 18

May 4. — (a) Plate X, No. 17. — S. branch of tail much the brighter.

(b) Plate X, No. 18. — All outpouring seems to be swept round to S. tail, even that from N.f. part of head. S. tail larger and brighter than N. No secondary nucleus.

May 4. — Plate III, No. 17. — Exposure from 14 h. 39 m. to 15 h. 09 m. Venus on plate. Tail 1° wide at 5° from nucleus. Very delicate and beautiful tail, slightly convex on South side. The tail is composed of a large number of overlying pencils or shafts. Running down the centre of the tail is a bright streak resembling a backbone. The North edge of the tail is simple, but from the South edge springs out a series of delicate branches at almost mathematically regular intervals. At about 6° 48′ from the head the tail bifurcates.

Transvaal Observ., Circular No. 4, 11th July 1910, p. 25, 28.

On MW 2, May 4.999, the structure of the head is very complicated. There are several jets of with two are most prominent. Jet *a* issues from the nucleus in p.a. 50° and turns back into the tail forming the northern branch of the inner (brighter) envelope. Jet *b* issues in p.a. 90° and joins the famer in the northern branch. The space between these two jets is also filled with luminous matter, so that the jets are only an increase of the local activity. It is impossible to detect any jets in the southern part of the envelope, but its edge es very bright and sharp and there is probably a continuous emission of matter into that branch. Both north and south branches fade rather rapidly into the tail at 5.′0 from the nucleus. Although there are many secondary ejections with a number of faint vertices on the sunward side of the inner envelope B, it has a rather sharp outline. The distance from the nucleus to the vertex is 27″ and the semilatus rectum is 54″, that is, the envelope is approximately parabolic. The phase-angle was 100°. There are also two very bright emissions symmetrical with respect to the axis of the tail, in p.a. 264° (p.a. of the prolonged radius vector was 256°). P.a. of these emissions are *c* 300° and *d* 228°. They extend only 1.′1, fading very rapidly, like jets *a* and *b*.

This is the first date on which two tails are recognizable. The first tail was sharp and straight; it usually consisted of a number of slender streamers. The second tail was of a diffuse nature and was usually more strongly curved and fainter than the first tail. I shall denote them as Tail I and Tail II, without implying that they were of Bredichin's first and second type.

The system of fine streamers constituting Tail I seems to originate in ejections *c* and *d*. As is true in most cases, these ejections from but the most conspicuous portion of a fan. There is a very strong northern streamer *C* issuing from jet *c* about 30″ from the nucleus. It is quite obvious that is does not begin at the nucleus proper. It is much the brightest among all the streamers near the head, but is does not extend very far from the nucleus. On WS 19, May 4.975, it can be seen only about 1°.4 from the nucleus, being very bright and sharply defined up to that point. There it fades away and becomes much fainter than the rest of the streamers. Apparently the emission in jet *c* was not continuous, and streamer *C* owes its peculiar brightness to a sudden and copious emission from *c*. Many other fine streamers are seen to originate at the tips of jets *c* and *d*, or between them.

It is interesting that this system of fine streamers intersects the southern (brighter) branch of the inner envelope without the slightest interference, which indicates that the two do not lie in the same plane. There is clear evidence of the motion of matter along the streamers, but there are no definite condensations for measurement.

Besides the jets described above, there is a bright halo, nearly circular, of diameter 3'.1, slightly displaced from the nucleus into the tail. There is also a very faint circular halo with the nucleus in its center. It has no definite boundary, but is quite noticeable for 4' on all sides of the nucleus. There are also traces of an outer envelope, especially on the north side.

MW 1. May 4.990, shows practically all the features spoken of. It is reproduced in Plate 12, figure 14, but all the details are lost in reproduction.

Crossley plates 2214, 2215, 2216 do not contribute anything essentially new. On C 2215, May 4.986, the streamers are especially well seen, due to a long exposure. They intersect very clearly the southern part of the envelope B and are seen to originate from jets *c* and *d*. The nucleus is approximately round, small, and slightly fuzzy on the sunward side. Its diamer is 6.'7 on MW 2.

On the Chile plate S7, May 4.892, reproduced in Plate 11, figure 13, the streamer Tail I and the fainter and brader Tail II are well seen. There is a condensation in Tail I just below the fork. It is marked on Plate 11, figure 13, with the letter A. The motion of this condensation will be studied later.

Lantern lens plates L3, May 4.975 and L4, May 5.974, show the two tails very distinctly. On May 4 the tails branch off at about 6° from the nucleus, but Tail I divides again into two branches at that point (the fork of Pl. 11, fig. 13). On May 5 this effect is even more striking. Tails I and II are widely separated and form a letter γ. However, owing to the presence of numerous lateral streamers, it would be difficult, if not altogether impossible, to trace the axes of the tails even near the nucleus.

There seem to have been rapid changes in the head about this time.

The observers at Johannesburg on May 3.625 note "slightly accentuated stream of superior brightness along S. branch of the tail" (Mitchell). Worsell and Innes describe the southern branch of the tail on May 4.625 as much the brighter. "All outpouring seems to be swept round to S. tail, even that from nf part of the head." Barnard, on May 3.931, observed the southern emission from the nucleus much brighter than the northern one. On May 4.927, the appendages were symmetrical.[42]

Quénisset[43] at Juvisy, on May 4.65 observed the head essentially the same as it was seen at Johannesburg on May 4.625, so that the northern outpouring developed between this date and 4.927. As we have seen, the MW and Lick plates show ejections distributed approximately uniformly north and south, and the northern side of the inner envelope B, although fainter than the southern, is comparable with it in brightness. The difference in brightness between the northern and southern sides of the envelope is overemphasized in the reproduction, Plate 12, figure 14.

On MW plates there are two or three curved jets in the *sf* quadrant from the inner envelope. They are much fainter than this envelope. They may be the remnants of the strong jets observed at Johannesburg, as their configuration is quite similar.

Objective prism spectrogram OP 6, May 4.896, shows four strong CO^+ tail images, and two fainter ones. These images are very strong for about 0°.5 from the nucleus, being much sharper and brighter on the violet (northern) side of the tail. This is due to the sharp northern edge of Tail I, described as a "backbone" (Wood, Johannesburg). The CO^+ streamers do not extend on the other (sunward) side of the nucleus. There is also some of the usual haze between the two blue CO^+ images. CN IV and CN III have jets *c* and *d* described above. The Swan knots do not show any emissions. The nuclear line is not clear owing to the blend with a star.

Objective prism W 205, May 4.977, shows confused images. There is a strong condensation in the far red, probably due to the Swan and red cyanogen bands.

Slipher's spectrogram and direct photograph of May 4, show that Tail I consisted exclusively of CO^+. The bifurcation and condensation A are well seen in CO^+. Slipher's conclusion is: "Comparison of this spectrogram with a simultaneous direct photograph identifies the tail emission streamers with the branches on the south side of the tail in the direct image and the yellow image of the tail with the more diffuse parts seen on the photograph." Evidently Tail I with all its branches consisted of CO^+ while Tail II gave the yellow image and continuous spectrum.

[42] See his drawings in Astroph. Journ. 39, 373, 1914.
[43] See his drawning. Bull. Soc. Astr. France, 24, 285, 1910.

N. T. Bobrovnikoff, Publ. of the Lick Observ., Vol. X VII, pt. 2, pp 351, 352, 1931.

May 05; observations see Fig. 2, No. 19 and 20

May 5. — Plate III, No. 18. — Exposure from 14 h. 46 m. to 15 h. 16 m. Venus on plate, Moon just off corner. Tail 1° 15′ wide at 5° from nucleus. The South edges of the envelopes surrounding the nucleus are brighter than the North edges. The tail consists of a number of filaments so numerous as to suggest a brush. Later the tail divides into three main branches. This branching gives the impression that the tail is concave towards to South.

May 5. — (a) Plate X, No. 19. — Nucleus much smaller, somewhat butterfly appearance. S. and N. branches of tail more even in luminosity.

(b) Plate X, No. 20. Nucleus small and round. Fan-shaped rays faintly seen. Brightest part of tail is the inner portion of S. branch, otherwise N. branch longer and generally brighter.

Transvaal Observ., Circular No. 4, 11th July 1910, p. 25, 28, 29.

Schweifabsätze waren meist nur angedeutet und konnten einigermaßen deutlich nur an einem Morgen, Mai 5, verfolgt werden und auch da nur, wenn das ganze Bild im Gesichtsfeld bewegt wurde. Ihre Länge an diesem Morgen betrug nur einige Bogenminuten, der Kern war verwaschen und oval, der innere Teil der Koma scharf abgegrenzt.

W. Münch, Astron. Nachr., Nr. 4425, S. 137, 1910.

MW 3, May 5.976, shows a strong jet *a* in p.a. 47° curving into the tail in the north side. There is also another jet b_1 in p.a. 78° curving in the same direction. Jet b_2 issues also in p.a. 78°, but curves in the opposite direction; that is, to the south. The length of all these jets is 70″. A very bright short jet *e* begins in p.a. 80° and sharply curves into the south side of the envelope, its vertex being 27″ from the nucleus. Jet *f* is shot out in p.a. 136°, but curves to the north side. These two jets *e* and *f* form a letter 0.

Jets *a* and *b* are undoubtedly those of the preceding date designated by the same letters. Jet *a* has changed its p.a. from 50° to 47°, and *b*, from 90° to 78°. The position angle of the prolonged radius vector has hardly changed at all (256°). The error of measurement in p.a. may be only a few degrees so that the change in p.a. of *b* seems to be real.

Jets *c* and *d* of the preceding date have lost their sharp outline, but are still recognizable. Their length is about 5′.

Envelope B of May 4 has considerably increased in size, but has preserved its parabolic proportions. It has its vertex 66″, semilatus rectum 132″ distant from the nucleus (instead of 27″ × 54″ als on May 4.999). This is due only in a small measure to the decrease in the geocentic distance (from 0.614 to 0.600). The velocity of expansion comes out to be 0.2 km/sec in the direction of the vertex and 0.4 km/sec in the direction of the latus rectum. It is of the same order as found for the expansion of halos. I think there is no doubt about the identification of envelope B. The configuration of May 4 and 5 is quite similar, the difference being chiefly in size.

It is more difficult to trace the origin of streamers than on the preceeding date, but many of them clearly originate in jets *c* and *d* and the space between them. They certainly do not originate directly from the nucleus. The faint outer envelope is obviously not parabolic. The nucleus appears triangular with sides of 5.′4 (MW 3). Crossley plates 2217, 18, 19 do not contribute anything essentially new. C 2218, May 5.983, is reproduced in Plate 12, figure 16. It shows a remarkable system of fine streamers. Each one of those visible on the reproduction can be divided in the original into a number of still finer ones. There is peculiar effect of mutual repulsion between individual streamers, well illustrated on this plate.

Although the presence of the diffuse Tail II is unquestionable, it has been impossible to trace its axis. In fact, the axes of the two tails seem to coincide. The halos have disappeared. They are not visible even on C 2218.

Drawings by Innes and by Worssell at Johannesburg on this date certainly illustrate the personal element in observing fine details. They differ not so much in the details themselves, but in the emphasis on them. Both drawings give approximately the same picture as Plate 12, Figure 15.

Condensation A, mentioned previously, was measured on seven dates in Tail I. Its was just below the fork on May 4.892 (see Pl. 11, fig. 13), but did not present any definite outline. Therefore the measurement is only approximate. Table 20 gives the results of measurement and reduction.

Tab. 20

Point	GMT May	1910.0 α	δ	R	w	r	v	Remarks
1	3.615	$23^h35^m10^s$	+8°24′	0.6726	+38°31′	0.6613	+39°27′	J 16
2	4.101	49 28	8 10	.6916	39 51	.6663	40 39	H
3	4.615	44 38	7 52	.7135	41 10	.6713	41 49	J 17
4	4.888	42 27	7 47	.7234	41 10	.6741	42 28	S 7
5	4.971	41 56	7 44	.7262	41 45	.6750	42 39	WF 19, WS 19
6	5.111	38 36	7 35	.7369	41 40	.6765	42 59	H
7	5.622	31 07	7 00	.7657	43 41	.6820	44 09	J 18

... By successive approximations we obtain: $(1 - \mu) = 27 \pm 3.6$.

... It is interesting that this orbit does not intersect the orbit of the nucleus. The closest approach to the nucleus occurred on May 2.5, but at a distance of 0.01 A.U. Another interesting fact is that the junction of streamers studied on

April 28 showed practically the same $(1 - \mu) = 23$. The identity of this junction with fork A is excluded since A passed perihelion on May 2.0. But it is probable that they both belong to the same phase of cometary activity on its south side.

N. T. Bobrovnikoff, Publ. of the Lick Observ., Vol. X VII, pt. 2, pp 333, 334, 1931.

May 06; observations see Fig. 2, No. 21 and 22

May 6. — (a) Plate X, No. 21. — Head decidedly yellowish, an appearance not noticed before. A star was involved in the tail and another was free; the mags. were estimated as 9.0 and 8.0 respectively, a difference of one magnitude. These were identified as B.D. + 8°, 5 and B.D. + 9°, 5 mags. 8.9 and 8.7 respectively, a difference of only 0.2. Hence the tail dimmed B.D. + 8°, 5 to the extent of 0.8 mag., equal to one-half of its light.

May 6. — Plate III, No. 19. — Exposure from 14 h. 14 m. to 15 h. 14 m. Width of tail 1° 15′ at 5° from head. Two envelopes surround the nucleus. These are sharper and brighter on the South edge, and form a sheath. Tail composed of a number of rays radiating from a point behind the nucleus. Irregular cloudy patch near South edge at 10° from head.

Transvaal Observ., Circular No. 4, 11th July 1910, p. 25, 29.

May 6, Alupka, Krim, 15^h. Kern scharf begrenzt, $2^m.5$, gerader hellgelber Schweif von 14° Länge, Richtung fast parallel mit ω, ι, γ Piscium. Koordinaten des Schweifendes $\alpha = 345^\circ.2$, $\delta = +4^\circ.8$.

K. Pokrowsky, Astron. Nachr., Nr. 4420, S. 59, 1910.

Plate 12, figure 17 represents S 8, May 6.901. Tail I is sharp and full of streamers. Tail II constitutes the continuous background. Tail I has a wedge shape. On the negative there is a distinct dark lane, the axis of symmetry. It is not quite dark but appears so rather by contrast.

There is a very faint jet forming the outer envelope, barely traceable and almost concealed by the halos. It begins in p.a. 135° on the south side and curves into the north side. Its maximum length is about 2′. This is probably jet f of the night before. A faint circular halo is visible on the MW plates. Its diameter is 5′. The nucleus is very small and seems to be elongated in the direction of the axis of the tail. Its diameter along this axis 2.′7.

Objective prism spectrogram OP 7, May 6.902, shows strong CO^+ bands $\lambda\lambda$ 4000, 4251, 4543 and fainter $\lambda\lambda$ 4381, 4686, 4150. On the direct photographs of this date Tail I has two bright streamers on both edges. These two streamers are well seen in the stronger CO^+ bands up to the edge of the plate, about 2°.3 from the nucleus. The structureless Tail II is well visible on the direct photographs on the south side (see Pl. 12, fig. 17) being much fainter than Tail I. At 2° from the nucleus it forms a protuberance (near the bright star on the right side of Pl. 12, fig. 17). This protuberance is well seen in the CO^+ images. This means that CO^+ was to some extent a constituent of both tails. This effect is better seen in the violet (south) edge of the images both because Tail II is brighter on that edge and because the CO^+ bands degrade to the red. Even the relative intensities of Tails I and II are preserved in the spectral images.

The CO^+ spectral images do not extend on the sunward side of the nucleus, in spite of the fact that the sunward side on the direct photographs is much the brighter.

The CN III_1 band has faint curved jets on the sunward side extending about 1′. They undoubtedly correspond to the curved jets *a* and *b* of this date. CN III_1 is considerably stronger and somewhat larger than CN III_2. There is some indication of jets *a* and *b* in CN IV, but they are concealed by the over-exposed head. Swan C IV and C V show these jets but very faintly, which is surprising since C IV in the head is very much brighter than CN III. CN IV_1 (and the much weaker CN IV_2) extends into the tail 1′ and shows the curved lateral jets *m* and *n*. The C IV heads are very weak in this region. The nuclear line is somewhat broadened by guiding. It probably corresponds to the nucleus proper and has in addition to CN and C_2 bands also C + H bands and H and K solar absorption lines.

Innes notes (Johannesburg, May 6.614): "Head decidedly yellowish, an appearance not noticed before. A star was involved in the tail and another was free; the magnitudes were estimated as 9.0 and 8.0 respectively, a difference of one magnitude. These were identified as BD. +8°.5 and BD. +9°.5, magnitudes 8.9 and 8.7 respectively, a difference of only 0.2. Hence the tail dimmed BD. +8°.5 to the extent 0.8 magnitude, equal to one-half of its light." It follows from this description that the star in the tail was about 10′ from the nucleus in the strong south side of the envelope.

The Johannesburg drawing 21 and 22 show well the "dark" axis and ejection to the south. One of them (No. 22) shows the beginning of ejection to the north, that is, jet *n*, but not jet *m*. Apparently *m* and *n* developed at about this time (May 6.61). The outer envelope on both drawings (in the original) is made only slightly fainter than the inner. The difference in brightness between these two envelopes on MW plates is striking. If seems, therefore, that the outer envelope was very weak photographically.

May 07

MW 8, May 7.002, shows two curved jets, a and b, in p.a. 75° and 114°. These may be jets *e* and *f* of the preceding night whose p.a.'s were 80° and 136°, but the distance from the nucleus to the vertex is again 27″ and the semi-latus rectum 54″. If these are jets *e* and *f*, it seems strange that they have not increased in size since the preceding night, as they usually do. Probably they are a new set of jets issuing from the same place as the night before. They form a half-envelope of parabolic proportions on the south side.

There are two short bright jets *m* and *n* in p.a. 165° and 335°, which are approximately symmetrical with respect to the axis of the tail. These two jets are well visible on the reproduction (like secondary nuclei) (Pl. 13, fig. 18). On the strongly exposed MW 6 and 7 all the jets blend together and give the head a symmetrical appearance. The tail is also symmetrical; the axis of symmetry being in p.a. 255°, that is, coinciding with the prolonged radius vector.

N. T. Bobrovnikoff, Publ. of the Lick Observ., Vol. X VII, pt. 2, pp 354, 355, 1931.

Mai 7 wurde er von Herrn Prof. Schwarzschild und mit ebenfalls zu etwa 6^m geschätzt und konnte der Komet leicht von beiden Beobachtern als rötlicher Stern mit bloßem Auge wahrgenommen werden.

W. Münch, Astron. Nachr., Nr. 4425, S. 138, 1910.

No. 4, T, May 7. — The tail, which is very faint towards the end, shows for about 27°.5. An irregular, thin, brighter strip begins 6° from the head. It is sinuous and apparently becomes the axis of the tail. From the head to this point the tail is uniform. About 9° from the head the south side of the tail is concave to the south and then becomes irregular. On the south side of the tail is a faint streamer which begins about 6° from the head and seems at its beginning to curve abruptly outward from the tail. Two faint diffuse masses appear near the end of the tail, the first is 17° and the second 20°.5 from the head. Towards its end, the tail has a tendency to curve northwards.

E. E. Barnard, Publ. Astron. and Astroph. Soc. Am., Vol. 2, p. 225, 1915

MW 9, May 7.987, shows such a complicated structure that it is impossible to describe it in detail. Apparently jets *m* and *n* of the previous date have expanded in all directions and started sending out a great number of fine streamers (about

1′ from the nucleus). Many streamers originate also on the sunward side of the envelopes. There is a bright halo of about 4′ diameter and a much fainter one of 12′. Their edges are extremely fuzzy. The nucleus is fuzzy and round, 6″—7″ in diameter.

... Especially prominent is the northern (left) edge of the envelope whose motion it is possible to follow for several days. Its sharp tip is marked with the letter *N*. ... It is evidently the expanded jet *n* of the previous date and the detailed investigation bears out this conclusion.

Table 24 gives the results of measurement of the tip of envelope *N* and reduction to the plane of the orbit of the nucleus.

Table 24

Point	GMT	1910.0		Envelope *N*				Remarks
	May	α	δ	*R*	*w*	*r*	*v*	
1	7.984	$0^h10^m11^s$	+ 9°38′	0.7090	+49°14′	0.7085	+49°22′	MW 9
2	8.572	12 41	9 52	.7154	50 16	.7154	50 36	HR 13, 14
3	8.984	14 19	10 03	.7199	50 51	.7205	51 27	MW 10
4	9.574	17 07	10 15	.7290	52 12	.7276	52 38	HR 15
5	9.967	18 49	10 24	.7348	52 55	.7325	53 25	WS 22, WF 22
6	10.563	22 00	10 42	.7423	53 53	.7399	54 34	HR 17
7	10.968	24 08	10 54	.7484	54 34	.7450	55 22	WF 23

It is evident that the motion in the envelope is in a straight line and that there is no appreciable acceleration. ... jet *n* has produced the northern wing of the envelope. It is remarkable that while expanding many hundreds of times jet *n* actually gained in brightness ...

Table 27

Point	GMT	A (1919.0)		B (1910.0)		Remarks
	May	α	δ	α	δ	
1	11.888	$0^h33^m31^s$	+11°43′	$0^h26^m25^s$	+11°20′	Cor; S 11
2	11.974	33 07	11 33	26 02	11 21	WS 24; WF 24
3	12.098	32 31	11 36	25 20	11 18	H
4	12.109	31 42	11 34	24 32	11 18	H
5	12.617	27 44	11 23	18 24	11 03	J 26
6	12.886	23 21	11 12	15 54	11 09	Cor.
7	13.104	19 43	10 52	9 21	10 48	H 13, 105; 13, 107

It was possible to measure in the tail two condensations A and B which might have been generated by the envelope *N*. The measurement is not very reliable in view of the diffuseness of these formation but it will suffice to show that the repulsive force acting on them was of an entirely different order of magnitude in comparison with that acting on envelope *N*. The results of measurement are:

. . . FORMATION A . . . $(1 - \mu) = 43 \pm 14$ . . .

Near May 9.8, tip *N* of the envelope and formation A would have occupied nearly the same place. Therefore the probability that condensation A was formed from tip *N* of the envelope is supported by direct evidence; this tip was forming streamers during the whole time of its existence.

There is yet another possibility. On May 11.888 (point 1A) tip *N* of the envelope and formation A were again nearly in contact. Thus A may have been formed only at that point. It is difficult to settle this question definitely as all the computed orbits are approximate.

It is quite possible that several contributing factors played their part in the formation of condensation A, namely: envelope near point 5N, nucleus near point 5N, nucleus near point 7N, envelope near point 1A. If such is the case some part of condensation A moved in front of the nucleus for about two days, that is, formed a jet toward the Sun. Phenomena like this may account for the extreme turbulence in the head when once-ejected masses again come into contact with the nucleus.

. . . FORMATION B . . . $(1 - \mu) = 27$ . . 5.0

. . . there is a close approach of condensation A and B . . . about May 10.7. B may have been formed at that point.

. . . envelope *N* on MW 9, May 7.987, sends out a great number of fine streamers. S 9, May 7.879, and S 10, May 7.913, show that these streamers do not go far into the tail. They extend only 3°.5 from the nucleus. Consequently they are a recent formation.

S 10 is reproduced in Plate 13, figure 20. Tail I and II are very distinct. Tail I is full of condensations but their measurement was impossible. There is a remarkable detached streamer on the south (right) side of Tail I, noticeable on the reproduction.

Comparison of Plate 12, figure 17 and Plate 13, figure 20 gives an impression of oscillatory motion of Tail I in reference to Tail II. In figure 17 both axes coincide, figure 20 the p.a. of the axis of Tail I is at least 3° to the south of the axis of Tail II. The change is probably due to the motion of matter in Tail I. The fork of the streamers moved at least 1°.5 between these two exposures. Precise measurement was impossible.

Spectrogram OP 8, May 7.896, has been reproduced and described by the writer.[44] It can be compared with the direct photograph S 9, May 7.879. On this photograph the brigth envelope (whose northern edge was denoted by *N*) has very sharp edges near the head. It extends toward the Sun for 4!6. The strong edges of this envelope give rise to jets at CN III_1 and CN III_2 on the spectrogram. The CN IV bands have such strong envelopes that the jets are entirely concealed by them. CN II_2 has also a faint jet.

The appearance of these emissions on the sunward side of the CN knots is of great interest. It shows that the bright envelope on May 7 and consequently jets *n* and *m* on May 6 consisted of cyanogen. Moreover, the CO^+ images, especially λ 4274, have an indentation on the red side corresponding in shape and dimensions to the northern branch of envelope *N*. Apparently all the envelopes consisted chiefly of CN, CO^+ being present also, but in comparatively small quantities. CN IV has a diameter of 7!5, which is probably due to the outermost halo (5′). The inner core measures about 2′, roughly corresponding to envelope *N*. There is an extremely faint image of the tail in CN IV about 1° long. It belongs apparently to the fine streamers issuing from tip *N* of the envelope.

The Swan C IV band is fully as strong as CN IV, but is much smaller. It corresponds to the inner core of CN IV. There are no sunward jets at the Swan bands. This is striking when we compare C V and the neighboring CN III, both of which are of the same intensity in the head. While CN III has remarkably bright jets, C V has not the faintest trace of them.

The CO^+ images do not go toward the Sun farther than the nucleus. This is worthy of notice, since envelope *N* is just as strong on the sunward side as away from the Sun. The details of the tail on the direct photographs can readily be identified in the CO^+ spectral images. The bifurcation of Tail I is well visible in CO^+. It is evident (as has been found previously) that both Tails I and II consist to some extent of CO^+.

This is one of very few spectrograms that extent into the ultraviolet. There is a faint knot at at approximately λ 3500, due to CN V. The spectra of the stars on the same plate are relatively much stronger in the ultra-violet than the nucleus.

This spectrogram furnishes direct proof that the envelopes of Halley's Comet on May 7 consisted chiefly of cyanogen. The CO^+ tail appeared as a by-product of some mysterious process of transformation. The CN envelope was issuing fine streamers, which later formed Tail I. While the motion in the envelope showed a repulsive force of the order 1, the motion in the tail gave $(1 - \mu)$ 43 and 27. This means that CN is subject to repulsion from the Sun in an entirely different way from CO^+.

[44] Publ. Astr. Soc. Pac. 40, Plate 29, 1928.

N. T. Bobrovnikoff, Publ. of the Lick Observ., Vol. X VII, pt. 2, p. 361—363, 1931.

Observations of Halley's Comet made on board s. s. "Mobile"
By Captain *P. Bisson* (Communicated by Capt. *Campbell Hepworth*)

On May 7, at about 3^h30^m a.m. mean time at ship, in approximate position lat. 39° 22′ N., long 59° 40′ W.

Observed angular length of comet's tail 7° 12′.

The night was dark and the comet showed up plainly. The head appeared about the brightness of a second magnitude star. Did not observe any split in the tail.

Meterological Office, London, S.W.; 1910 June 6.

Monthly Notices of R.A.S., L XX, 8 (1910) 617.

May 08; observations see Fig. 2, No. 23 and 24

Mai 8. Comme le 3 Mai, le noyau parait double, divise en deux plages brillantes séparées par une partie sombre. — Mai 15. Noyau 12″, éclat 0.4 gr. à l'oeil nu.

F. Gonnessiat, Astron. Nachr., Nr. 4453, p. 205, 1911.

The tail is 25° long, uniform and diffuse and slightly diverging. No detail of structure. There seems to be a faint irregular branch on the south side of the tail, beginning 9°.5 from the head.

E. E. Barnard, Publ. Astron. Astroph. Soc. Am., Vol. 2, p. 225, 1915.

May 8. — (a) Plate X, No. 23. — Taken altogether, the S. branch is the brighter. Curious gap between the two main jets. S. edge of the brighter jet was along 85°; it was slightly curved. Nucleus small and stellar.

(b) Plate X, No. 24. — No secondary nucleus. Very prominent gap in fan of streamers.

May 8. Plate III, No. 20. — Exposure from 15 h. 08 m. to 15 h. 23 m. The earlier part of this morning was spent in photographing the end of the tail. The impression was very faint, and no detail could be seen. Tail 45′ wide at 5° from nucleus. Inclination of main axis of tail to meridian 258° from North through East. Tail straight with no branches: composed of a number of rays, ten being visible near the head.

Transvaal Observ., Circular No. 4, 11th July 1910, p. 25, 29

MW 10, May 8.987, . . . shows clearly the expanded envelope *N*. The peculiar configuration of May 7 was fully preserved and there is no difficulty in the identification. There are clear signs of exhaustion of matter moving from the former jet *n* (May 6); edge *N* of the envelope is sharply defined on its lower side. Emission has been in progress for several days and has now stopped. The lower edge of the envelope is the last ejection of matter from jet *n*.

There is also an inner envelope well developed on the north side.

The outer envelope *N* is brighter on the north side. Some of the fine streamers intersect it without any disturbance. Others are deflected and much weakened by this envelope. It is evident that at least some of the streamers must lie near the plane of the orbit of the nucleus.

Helwan HR 13, May 8.572, shows two bright slightly curved jets *p* and *q* in p.a. 96° and 136°; *p* is 60″ long, *q* is only slightly shorter. In MW 10, May 8.987, the p.a.'s are 91° and 141°. *p* is 81″ long and *q* became a quite indefinite, broad je 50″ in length. It is possible to recognize *p* on MW 13, May 9.988, in p.a. 94°; its length is 151″, so that there seems to be a considerable acceleration of motion toward the Sun.

Most of this acceleration comes from the diminishing geocentric distance. The vertex of jet *p* is approximately in the radius vector (p.a. of which was 75°). From this we find that the average velocity between May 8.572 and May 8.987 was 0.21 km/sec; while between May 8.987 and May 9.988 it was 0.27 km/sec. There seems to be no doubt that the motion was slightly accelerated. At least there was no appreciable retardation as might be expected. This circumstance will be used for the determination of the mass of the comet. Jet *p* can be seen in figures 21, 22, 23, pointed downwards.

MW 10, May 8.987, shows jets *r* and *s* 75″ in length. The velocity of expansion is, therefore, of the same magnitude as in jet *p*. New features on MW 10 are: the development of a broad and very brigth fan in p.a. 321° to 91°, 43″ long, in which several jets can be distinguished; these and the previous formation constitute a complicated structure with a maze of details. The size of the outer envelope *N* (perpendicular to the axis) is on the Helwan plate HR 14, May 8.578, 2′06″ and on MW 10, May 8.987, 2′47″. It is impossible to distinguish the vertex in a network of jets. The details of the branch *N* of the envelope were so well preserved for several days that there is no doubt of the fact of the expansion of the envelopes with a velocity approximately 0.3 km/sec.

The envelopes on all plates of this date show sharp edges. As many as five envelopes can be seen on MW 10 (Pl. 13, fig. 21), on the north side. Although the space between them is also filled with luminous matter, the discontinuity in brightness is very striking. This gives additional evidence that the process of ejection occurred intermittently.

The nucleus is 5″ in diameter, sharp and bright (MW 11).

N. T. Bobrovnikoff, Publ. of the Lick Observ., Vol. XII, pt. 2, p. 363, 1931.

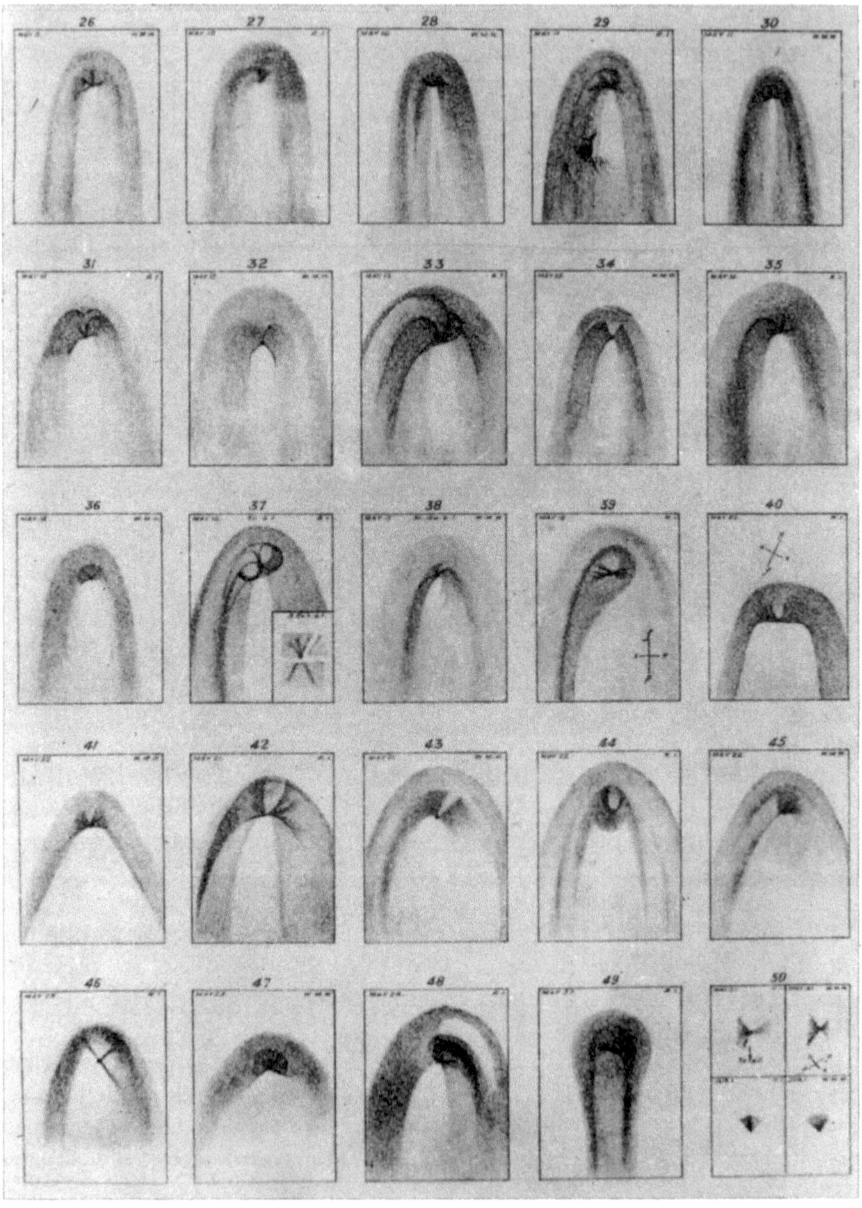

Fig. 3. Evolution of Halley's comet between May 09 and May 31, 1910; drawings taken from R.T.A. Innes; Transvaal Observatory Circular No. 4, 11 th July 1910, Plate XI.

May 09; observations see Fig. 2, No. 25 and Fig. 3, No. 26

Exposure from 14 h. 17 m. to 14 h. 48 m. Tail 1° 9′ wide at 5° from nucleus. Inclination to meridian 256°. Tail mainly straight, with a bright streamer leaving the South edge at ablut 1° 30′ from the nucleus. The tail contains a large number of rays, the two brightest being near the North edge. The nucleus of the comet is now very bright, and the envelopes appear to continue and form a sheath or cap over the tail.

May 9. — Plate IVm No. 22. — Exposure from 14 h. 53 m. to 15 h. 24 m. Tail 1° wide at 5° from nucleus. Inclination to meridian 257°. Details similar to No. 21. Between the streamer leaving South edge of tail and the main tail appear two fainter streamers. Long bright sheath.

May 9. — (a) Plate X, No. 25. — Nucleus somewhat larger. N. branch of tail near head is the brighter.

(b) Plate XI, No. 26. — No secondary nucleus. N. branch of tail brighter near head. S. f. boundary of head ill-defined. Traces of an inner heath in N. branch.

Transvaal Observ., Circular No. 4, 11th July, 1910, p. 25, 29.

May 9. The main tail is 31° long. It diverges slightly and soon becomes very faint and diffuse. Along the axis is a rather narrow diffuse streak, which cannot be traced nearer to the head than 4°. A number of thin faint streamers are visible on the south side of the tail, along with some irregular markings.

E. E. Bernard, Publ. Astr. and Astroph. Soc. Am., Vol. 2, p. 225, 1915.

La tête de la comète est composée d'un petit noyau un peu flou, mais bien apparant, situé à l'intersection de deux aigrettes opposées par le sommet et perpendiculaires au mouvement diurne. Légère enveloppe gazeuse, un peu plus accentuée de coté du levant. La 3ᵉ série de mesures est faite alors que la comète est à la limite de visibilité. A la lumière croissante de l'aurore, la comète est disparue que l'on distingue encore 1′ 35 cataloquée de grandeur $7^{m}_{\cdot}3$.

P. Chofardet, Astron. Nachr., Nr. 4489, S. 4, 1911.

Am 9. Mai zeigte der Ausströmungskegel drei zonenartige Unterbrechungen senkrecht zu seiner Achse.

M. Wolf, Astron. Nachr., Nr. 4416, S. 402, 1910.

Mai 9. Komet heute auffallend schwach, im Sucher schwer sichtbar. Kopfhelligkeit kaum 3^m—4^m. Schweif im Hauptrohr nicht sicher erkennbar. Der sichtbare Teil des Kopfes hat 1′ im Durchmesser und ist stark gelb gefärbt. Kernhelligkeit etwa gleich dem Stern + 10°31 (PD $7^m_.26$).

K. Graff, Astron. Nachr., Nr. 4539, S. 46, 1912.

MW 14, May 9.994, shows a number of short and bright jets. Jets *a*, in p.a. 19°, length 30″ and *b*, in p.a. 64°, length 38″, are almost straight. Jet *c*, im pa. 94°, length 40″, is perfectly straight up to 27″ from the nucleus; at that point is sends out a curved and much fainter jet, thus forming a secondary center of emission. Jet *d*, in p.a. 136°, lenght 27″, is straight and sharp. Jet *e*, in pa. 239°, length 41″, is very fine and slender. It exists only on this plate: MW 12, May 9.980 and MW 13, May 9.988, do not show it at all. There is another jet *f*, in p.a. 279°, of the same length and aspect as jet *e*, making a fine pair. Since the position of the prolonged radius vector for this date is p.a. 255°, jets *f* and *e* are situated on both sides of this line. Jet *g* forms a pair with *a*, its p.a. 259°, but it is less definite than *a*. The long jet *p* has already been mentioned. There are several semienvelopes on the north side corresponding to the predominant direction of emission . . .

The nucleus is round and very bright. Its diameter is 5″.

N. T. Bobrovnikoff, Publ. of the Lick Observ., Vol. X VII, pt. 2, p. 364, 1931.

May 10; observations see Fig. 3, No. 27 and 28

May 10. — (a) Plate XI, No. 27. — Nucleus larger. (b) Plate XI, No. 28. — Nucleus slightly elliptical, larger and less well-defined. S. branch of tail brighter and broader for more than a degree from head, slight nebulosity outside principal sheath.

May 10. — Plate IVm No. 23. — Exposure from 14 h. 19 m. to 14 h. 49 m. Inclination of tail to meridian 257°. The tail appears slightly convex towards South. The edges of the envelopes are sharper on the North Side than on the South, otherwise the emission appeared symmetrical. The tail is not so bright along the centre line.

Transvaal Observ., Circular No. 4, 11th July 1910, p. 25, 29.

Most of the observations were made with black wires against a bright sky. The heavens were very clear, and the comet's tail must have been 20° long, broadening beautifully as it receded from the head; it seemed perfectly straight. The head was bright, and nearly filled the field of view in the 20 inch, which was 15′ in diameter. There was a dark space right behind the nucleus.

H. A. Howe, Astron. Nachr., Nr. 4784, S. 134, 135, 1912.

MW 17, May 10.993, has a remarkable system of curved jets, of which only a faint idea can be gained from Plate 14, figure 25 (MW 15, May 10.982). The position angle and length of the jets is given in Table 38, together with the data for May 9. The identification of various jets is by no means certain, but the general agreement is sufficient to show that there was no appreciable rotation.

Table 38

Jet	May 9	May 10	Change in p.a.	Length		Change in length
				May 9	May 10	
a	19°	27°	+ 8°	30″	81″	+51″
b	64	61	− 3	38	68	+30
c	94	125	+31	40	46	+ 6
d	136	149	+13	27	67	+40
e	239	233	− 6	41	67	+26
g	359	336	−23	30	53	+23

The general increase in length is chiefly due to the decreased geocentric distance. The velocity of expansion would be of the order of 0.2 km/sec. Jets *a*, *b*, *c*, *d*, *e*, on May 10, all curve to the south side; only *g* curves to the north. There is no trace of jet *f* of the preceding date. The nucleus is fuzzy and from 6″ to 7″ in diameter.

C 2225, May 10.975, is reproduced in Plate 14, figure 26. It is remarkable for the complex structure of its envelopes, especially on the north (left) side. The other side is diffuse. The dark axis of symmetry is well shown.

Objective prism spectrogram W 211, May 10.979, shows long tail images overlapping each other. CN IV has a strong emission toward the Sun about 2′ in length, which roughly corresponds to the sunward jets *b*, *c*. C II, C III and C IV are very strong and likewise show this emission. W 212, May 10.979, has the same ejections toward the Sun. CN IV has a faint halo or envelope extending towards the Sun 2′.5; it is probably the outer envelope. C IV has as usual a very bright tail, but neither C III nor C II show any, although C III is fully as bright as C IV. The tail at C IV is probably due to reflected light, but traces of CO^+ bands are easily recognizable. There is a diffuse tail near λ 7000.

N. T. Bobrovnikoff, Publ. of the Lick Observ., Vol. XVII, pt. 2, p. 364, 1931.

Mai 11; observations see Fig. 3, No. 29 and 30

No. 7, T, May 11. The tail is 35° long. It widens and is very faint toward the end where its width is about 4°.3. There is little or no structure in it. A narrow, dark streak or line runs along the axis for a distance of about 6°. This apparently continues farther, about the same distance as a bright line.

E. E. Barnard, Astron. and Astroph. Soc. Am., Vol. 2, p. 225, 1915.

May 11. — (a) Plate XI, No. 29. — From this date until the day of transit, the comet with its enormous bright tail was an imposing and beautiful spectacle, far transcending the comets of 1881 or the Great Southern Comet of 1901.

(b) Plate XI, No. 30. — Nucleus elongated, streamers not very prominent. Principal sheath more sharply defined on N. side; on S. side it melts into the outer glow.

Exposure from 14 h. 33 m. to 15 h. 06 m. Inclination of tail to meridian 257°. The centre of the tail is dark, with a bright ray running along the North side of the dark space. This ray curves towards the South at 4° from the head.

May 11. — Plate V, No. 25. — Exposure from 15 h. 16 m. to 15 h. 28 m. Two sharp envelopes persist for about $1^1/_2$° behind nucleus, forming a sheath for the tail. A long dark lane behind the nucleus divides the tail into two symmetrical parts. The Northern half is bounded by a straight bright ray along the inner edge. Inclination of tail to meridian 257°.

Transvaal Observ., Circular No. 4, 11th. July 1910, p. 25, 29.

May 11.976, has strong emission in p.a. 72° and 150°, probably corresponding to jets *b* and *d* of May 9 and 10. There is also a fuzzy jet in p.a. 241°, which is probably *e*. The relative intensity of the jets is approximately the same as on May 10. Remnants of all the other jets of May 10 can also be distinguished.

The nucleus is large and fuzzy; its diameter is 3″—4″.

Objective prism spectrogram OP 10, May 11.891, shows C IV very much brighter than CN IV, not only in the core but also in the envelope; this is the reverse of what has been observed before. C V is also much brighter than CN III. Probably CN diminished in brightness very considerably. Both C IV and CN IV are better developed on the sunward side, which corresponds to jets *b* and *d*.

C2230, May 11.976, is reproduced in Plate 15, figure 28. The details near the nucleus are entirely concealed by a general glow. The striking feature is the dark lane along the radius vector. It can also be seen in figure 27 of Plate 14, S 11, May 11.894. At about 2° from the nucleus it deviates very considerably from the prolonged radius vector and from the axis of symmetry of the tail and becomes quite irregular. This is the best proof that it is not the shadow of the nucleus but is simply due to the comparative scarcity of emission in that direction.

N. T. Bobrovnikoff, Publ. of the Lick Observ., Vol. X VII, pt. 2, p. 368, 1931.

May 12; observations see Fig. 3, Nov. 31 and 32

1910 Mai 12 bot er am kleinen Refraktor (Öffnung 16 cm) bei 50facher Vergrößerung ein sehr schönes Bild dar. Die phasenartige, gut begrenzte große Kernscheibe zeigte sich in goldgelber Farbe (5^c); der Schweif ließ sich auf dem hellen Dämmerhimmel zwar nur ein kurzes Stück verfolgen, doch waren die von der Kopfpartie beiderseits ausströmenden, von der Sonne weg gerichteten Ausläufer gleich der ausgedehnten, den Kern im Halbkreis umgebenden Koma sehr detailliert gezeichnet.

Mai 12. M.Z. Strb.	Total-helligkeit	Red. auf $r = 1, \Delta = 1$	Reduz. Helligk.	Bemerkung
$15^h.5$	$-0^m.3$	$+3^m.0$	$2^m.7$	Unsicher, ZD = 80°

C. Wirtz, Astron. Nachr., Nr. 4446, S. 93/94, 1911.

. . . Die Aufnahme zeigt, daß am 12. Mai eine ausgedehnte Wolke von Kometenmaterie dem visuellen Schweif nördlich etwas voranflog. In der Skizze ist zur besseren Vergleichung der visuelle Schweif nach meiner eigenen nahe gleichzeitigen Beobachtung eingezeichnet.

Es ist wahrscheinlich, daß auch an den folgenden Tagen ähnliche Wolken von dem Kometen ausgegeben wurden. Bei der Diskussion der Frage, ob die Erde durch die Schweifmaterie gegangen ist, müßten daher diese Wolkenbildungen mit in Betracht gezogen werden.

Königstuhl-Sternwarte, Heidelberg, 1910 Juni 2.

M. Wolf, Astron. Nachr., Nr. 4421, S. 69/70, 1910.

Um 14^h m. Z. Bamberg vor Aufgang des Kerns konnte der gerade Schweif bis zwischen α Aquarii und ϑ Pegasi gut verfolgt werden; er war also über 40° lang (43 Millionen km). Um 15^h Kopf um $1^m.5$ schwächer als der in gleicher Höhe und auf gleich hellem Grunde stehende α Aurigae, also 2^m. Kern $0^m.5$ heller als 58 Piscium (PD $5^m.67$). Um den Kern eine prachtvoll rötlich-goldene Hülle (2 Büschel).

E. Hartwig, Astron. Nachr., Nr. 4413, S. 349/350, 1910.

May 12. — (a) Plate XI, No. 31. — Nucleus and bright coma decidedly yellow = 3. Extraordinary appearance. Jets are of Prince of Wales' Feathers type. Kink on S. side if dark space behind nucleus.

(b) Plate XI, No. 32. — Nucleus small, round, and brighter. Nucleus and streamers yellow = 3. Extraordinary ray from nucleus bent near end towards N.f. and terminated in a knob. Inner sheath well-defined on N. side, but melting into outer sheath at head and on S. side. S. tail brighter.

May 12. — Plate V. No. 26. — Exposure from 14 h. 27 m. to 15 h. 16 m. Tail 1° 15′ wide at 5° from nucleus.: Inclination to meridian256°. Envelopes surrounding nucleus extrend clearly for 1°, forming a sheath. The tail consists of a number of rays, apparently shooting out from behind the nucleus and gently curving outwards. There are two groups of these rays, the groups separating and leaving a dark division along the axis of the tail.

Transvaal Observ., Circular No. 4, 11th July 1910, p. 25/29.

No. 8, T. May 12. The tail is about 43°.5 long, but very faint for the last 12°. Near its end it is 3°. I wide. A thin rather irregular and diffuse bright streak runs along the axis of the tail to a distance of 10° or 11°, where it suddenly curves northward for 1° and then forms the northern boundary of the rest of the tail. The portion of the tail south of this line beyond the curve is uniformly denser.

E. E. Barnard, Publ. of the Astron. and Astrophys. Soc. Am., Vol. 2, S. 225, 1915.

A new phenomenon took place on May 12. The formerly diffuse glow about the head developed into definite edges A, B, C, shown on Plate 15, figure 29 (C 2235, May 12.982). It is seen from figure 30 of the same plate (C 2236, May 13.978) that these edges were moving into the tail. It was possible to follow their motion for nearly four days. Table 39 gives the result of measurement.

. . . The remarkable phenomenon which I have described here as the motion of formations A, B, C developed apparently very suddenly. Evershed does not mention it at all on May 12.475. The Kodaikanal photograph is reproduced in the *Monthly Notices* **70**, Plate 20, 1910, and is described as "noteworthy for the

Table 39

Point	GMT May	A α	A δ	B 1910.0 α	B δ	C α	C δ	Remarks
1	12.965	$0^h48^m50^s$	+12°26′	$0^h48^m45^s$	+12°25′	$0^h48^m48^s$	+12°28′	C 2234
2	13.973	1 02^m59	13 21	1 02 25	13 14	1 02 26	13 22	WS 26
3	14.572	12 31	14 00	12 08	13 51	12^m15	14 07	HR 20
4	14.623	13 20	14 03	12 49	13 55	13 20	14 15	J 28
5	14.977	20 19	14 21	19 41	14 17	20 19	14 36	WS 27
6	15.113	23 10	14 34	22 38	14 25	23 00	14 48	H
7	15.634	34 26	15 16	33 30	15 08	34 20	15 34	J 29
8	16.640	2 03 06	16 43	2 03 26	16 33	2 01 06	16 59	J 30

welldefined edge of the coma on the north side", that is, on the side of formation C. The Johannesburg observers on this night (May 12.619) say that the nucleus and coma was "decidedly yellow = 3. Extraordinary appearance. Jets are of Prince of Wales' Feathers type. Kink on S. side of dark space behind nucleus" — Innes. "Nucleus small, round and brighter. Nucleus and streamers yellow = 3. Extraordinary ray from nucleus bent near end toward nf. and terminated in a knob" — Worsell.

Figure 29 of Plate 15, represents C 2235, May 12.982 and figure 30 of the same plate, C 2236, May 13.978. These reproductions show only the outlines of formations A, B, C. The fine details are entirely concealed by envelopes. The increase in size is largely due to real expansion; p did not change much (from 0,36 to 0.32).

The kink on the south side referred to by Innes is due to the angle formed by jets A and B. Apparently they were not separated at that time. On C 2235 they are neatly separated; the distance between the lower end of A and upper end of B 12″. The "dark space" is dark in relation to the more brilliant parts of the head.

On the Crossley plates of May 12, the tips of both A and B show fine streamers extending into the tail. They can be traced on WF 25, May 12.970, to about 40′ from the nucleus. On WS 26, May 13.975, they are visible at least 1°30′ from the nucleus. The motion of matter in these streamers is evidently much more rapid than in their parent jets. In other words, $(1 - \mu)$ is greater in the streamers than in the jets.

An extensive streamer system on the north side can also be traced to tip C.

The structure of the head is exceedingly complicated. Besides formations A, B, C there are many interlacing jets. There is a definite condensation in p.a. 131° which is apparently moving toward the Sun. The distance between its center and the primary nucleus is, on C 2234, May 12.967, 18″, and on C 2235, May 12.982, 97″. This is probably what Innes describes as "a great piece thrown off," giving its distance from the nucleus as 40″ (in the original journal of observations). It is shown on both drawings in p.a. approximately 90° which throws some doubt on the identification. On Innes' drawing there is also another jet of approximately the same length but in pa.a. 135° which may be the one observed on the Crossley plates. It is hardly safe to deduce anything from this description except the fact that there was a rather rapid motion of matter near the nucleus, probably with a velocity in the neighborhood of 0.5 km/sec.

On the Crossley plates there is also a fan-shaped emission 62″ in length, the edges of which are in p.a. 31° and 71°. Its edge in p.a. 31° formed jet C. The brightness of various parts is, on an arbitrary scale, A 10, B 2, fan (and C) 8. The nucleus is ill-defined but there is a brighter part of about 7″ diameter (C2234).

A visual observation of the spectrum by Wright, May 13.0, emphasizes the unusual brightness of the continuous spectrum of the coma on the following (sunward) side, while the preceding side was relatively faint. The continuous spectrum of the nucleus was strong with the three usual Swan bands, C IV, C III, C II. The two latter could be followed into the tail only 15′. On the sunward side the Swan bands were brighter but shorter than on the tail side. The width of the continuous spectrum of the coma was 2′.1.

This observation (accompanied by a drawing) is fully confirmed by spectrogram W 217a, May 12.975. On this spectrogram, CN IV is much brighter but shorter on the sunward side, just as the Swan bands are.

The peculiar aspect of the continuous spectrum of the nucleus on this date is well illustrated by the MW spectrogram β 172 of May 11.993, that is, a day earlier (Pl. 21, fig. 65). The conditions in the head on May 12 and 13 were approximately the same. The continuous spectrum is evidently given by the sunward (upper in the reproduction) part of the head only. The peculiar feature of this spectrogram is the almost complete absence of the cyanogen bands. This can hardly be explained by the instrumental conditions, as the Fe comparison spectrum is approximately the same in the distribution of intensity as on April 24 (Pl. 21, fig. 63) and April 29 (Pl. 21, feg. 64), both of which have very bright cyanogen bands. Furthermore W 210, May 10.988 (Pl. 21, fig. 67) also falls to show the CN bands, although the focus was set nearer CNIV than on April 28 and April 20 (Pl. 21, fig. 66).

While there can be no assurance of the correctness of these conclusions because of the lack of observational data there is a circumstance which makes it very probable. On May 23, before a similar outburst in the head, the CN bands disappeared in the same way, as we shall see later. It seems that unusual activity of the nucleus is connected somehow with the repression of the cyanogen spectrum and of the violet end of the spectrum in general.

N. T. Bobrovnikoff, Publ. of the Lick Observ., Vol. X VII, pt. 2, p. 368/369, 1931.

May 13; observations see Fig. 3, No. 33 and 34

May 13. — (a) Plate XI, No. 33. — Nucleus small and quite stellar. S. side of tail brighter near head. Nucleus like a 4th mag. star and yellow. (14 h 45 m. G.T.)

Nucleus is throwing out an extension in 230° 1″, giving it a double-star appearance (15 h. 5 m. G.T.). Secondary is now at 240° $1^{1}/_{2}$″ (15 h. 23 m. G.T.). Triple appearance. Nucleus certainly grown in size since 14 h. 45 m. G.T. (15 h. 32 m. G.T. dawn.)

(b) Plate XI, No. 34. — Nucleus small, round, and well-defined, yellow = 3. Curious gap in coma N.f. nucleus. S. tail is the brighter.

May 13. — Plate V, No. 27. — Exposure from 14 h. 48 m. to 15 h. 28 m. Tail 2° wide at 5° from head. Inclination to meridian 258°. This photograph shows the envelopes brighter and more pronounced on the South side. Sheath structure well marked. The tail does not divide, but is fainter down the centre. A streamer passing out on the South edge of the tail shows a sharp bend.

Transvaal Observ., Circular No. 4, 11th July 1910, p. 26, 29.

On May 13 a new and remarkable phenomenon developed in the head: two entirely new jets *a* and *c*, slightly curved in opposite directions and "looking like the two wings of an insect," (Curtis). Jet *a* is in p.a. 259° and is 68″ long (C 2236 May 13. 978). Jet *c* is in p.a. 317° and is 87″ long. The p.a. of the prolonged radius vector was 257°, so that jet *a* coincided approximately with it in direction.

Both jets show condensations and have jagged edges. Apparently the explosion went on intermittently. There is also a well marked rift between the nucleus and these jets at about 10″ from the nucleus. At this place jet *c* is connected with jet *a* by a fine thread.

This phenomenon developed almost under the eyes of the Transvaal observers. Innes writes (May 13.628): "Nucleus is throwing out an extension in 230°, 1″, giving it a double star appearance." At 13.641, that is, 18^{m} later, the secondary nucleus was in p.a. 240°, $1^{1}/_{2}$″; at 13.647, that is, 9^{m} later, the nucleus was triple and had "certainly grown in size" since the beginning of observation. From this description and from the original drawings it follows that Innes observed the initial stage of the formation of jets *a* and *c*. Unfortunately this region is very deficient in comparison stars and it was not possible to measure the position of the tips of jets *a* and *c*. But since a was approximately on the prolonged radius vector it is possible to compute the value of the repulsive force. ...

The motion of matter was accelerated in reference both to the comet and to the Sun. The cometocentric and heliocentric radial velocities were found as follows:

The first value of the cometocentric velocity represents probably the velocity of ejection. It is surprisingly low; even if Innes's measurements are much in error the velocity of ejection cannot have exceeded 0.2 km/sec. ...

Table 49

First Observation	Velocity		Cometocentric	Remarks
	Second Observation	Heliocentric		
May 13 15^h05^m	15^h23^m	23.2 km/sec	0.10 km/sec	First abs. by Innes
May 13.628	13.978	24.0	0.64	Points (1) and (2)
May 14.574	14.980	26.5	2.85	Points (4) and (5)

Jets *a* and *c* did not constitute all the interesting features of the head on that night. There was a fan-chaped emission with the flanks directly opposite to the jets, that is in p.a. 79° and 137°, 26″ long. It was almost as bright as the nucleus itself. There was a curious, relatively dark, round formation with the center 113″ from the nucleus in p.a. 59°. It was rather sharply bounded on all sides and looks as if a ball of luminous matter expanded into a ring leaving very little in the center. There was no trace of the secondary nucleus of the preceding date. Besides the jets described there were several partly interlocking envelopes and ejections.

Both Johannesburg observers mention the deep yellow color of the nucleus on this date. Worsell describes the gap in the envelopes (studied above) but unfortunately gives no measurement.

The nucleus at the beginning of the Johannesburg observations was stellar but grew streadily in size. On the Crossley plates it is by no means stellar; it is diffuse with the brightest part 4″ in diameter (C 2236).

It is worthy of notice that the Johannesburg observers' drawings are very similar to the crossley photographs. Father E. Goetz (Bulawayo Obs., ibid.) also mentions the yellow color of the nucleus. From his drawing we can conclude that formations A, B, C of May 12—16 (which moved under $(1 - \mu) = 1$) were well visible to the eye and appeared yellowish; consequently they cannot have consisted entirely of cyanogen.

N. T. Bobrovnikoff, Publ. of the Lic O Observ., Vol. XVII, pt. 2, p. 369/372, 1931.

May 14; observations see Fig. 3, No. 35 and 36

No. 9, T. May 14. The tail is 50° long, but very faint for the last 12°. Near the end it is 4°. 3 wide. It is a little darker along the axis for some 9°, but it appears flat and without any structure. The head is very bright, with a bright portion extending into the tail on the south side.

E. E. Barnard, Publ. Astr. and Astrophy. Soc. Am., Vol. 2, p 225, 1915.

The fan-shaped ejection of the Crossley plates of May 13 constitutes the principal feature on the Helwan plate HR 20, May 14.574. It hat expanded very considerably and its edges are now in p.a. 45° and 170°. Although the opening of the fan is now 125° instead of the 58° of the day before, the p.a. of the axis of the fan has not changed. In other words the fan has opened up on both sides uniformly. The position angle of the prolonged radius vector did not change appreciably during this interval. The fan is now 90″ long in its axis but only 48″ at the edges. This again shows that the ejection of matter on both edges occurred more recently than in the middle. According to the Johannesburg observers there were several distinct jets in this fan. The head again appeared yellow.

Quénisset at Juvisy mentions[46] a rapid change in the head which occurred within 19^m. According to this drawing made on May 14.635 the nucleus had two broad emissions which look like the fan in the Helwan plates. On May 14.648 there were four fine jets as if the original two each split into two new ones.

The four Crossley plates of May 14 show that the fan opened up still more. It now occupies fully 180°. Its length along its axis is 85″ which is approximately the same as on the Helwan plates. There are several fine jets in the fan, whose p.a. it is impossible to measure because of the lack of comparison stars. Jet *c* of May 13 became so fuzzy that only its traces can be seen. There is no its diameter is only 20″ which shows how much the length of exposure affects the size of the nucleus. There is a definite sharp nucleus on the Helwan plates, so that this increase in size is a new development. The rear of the nucleus is entirely free of emissions except for the very faint receding jets *a* and *c* of May 13.

A rather faint spectrogram W 218a, Mai 14.984, shows a sharp nuclear line about 30″ in width and the usual emissions of CN and Swan. The CN IV knot gives a faint extension about 2′ long toward the Sun. Comparison with C 2242 shows that the central globular condensation had a continuous spectrum while the fan consisted chiefly of CN and C_2.

[46] Bull. Astr. Soc. France, 24, 285, 1910

N. T. Bobrovnikoff, Publ. of the Lick Observ., Vol. XVII, pt. 2, p. 373/374, 1931.

May 15; observations see Fig. 3, No. 37 and 38

(b) Plate XI, No. 36.

May 15. — (a) Plate XI, No. 37. — Unusual appearance. Most glow behind nucleus. (15 h. G.T.)

Nucleus changed, well-marked rifts. (15 h. 43 m. G.T.)

(b) Plate XI, No. 38. — Nucleus larger, elongated p. and f., yellow = 3. Streamers almost all behind a N. and S. line through centre of nucleus. (15 h. 10 m. G.T.)

N.B. — The comparison-star used this morning was in the coma and was estimated as 9.5, B.D. gives 8.8. It is quite obvious that a star behind the brillant coma must appear dimmed. The statements on this point in some popular books are inexact.

May 15. — Plate VI, No. 29. — Exposure from 15 h. 00 m. to 15 h. 28 m. Longest photographed tail. Very slightly convex on South side. Inclination to meridian 259°. Four envelopes visible, with sharper edges on North side, but brighter on South side. Long dark region along axis of tail.

Transvaal Observ., Circular No. 4, 11th July 1910, p. 26/29.

On May 15 and 16 Dr. Curtis exposed small-scale plates on various portions of the tail. His description of the tail is of interest and will be given here verbatim (with some changes in notation).

"Plates WS 28 and WF 27 (both May 15.948) were taken on the tail at 40° from the head, guiding on a star. They show a very broad structureless tail. At this point the tails is 4° in total breadth, made up of a wide northern branch 2°.5 wide and a narrower southern branch 1° wide. 58 Pegasi is nearly 1° south of the northern branch of the tail; there are traces of a still fainter southern streamer passing about 30′ north of θ Piscium. Counting this streamer, the tail is 5° broad at 40° from the head."

"Z 6 (May 15.948) shows the region of the tail in Aquarius at 60° from the head. Tail broad and structureless as fas a individual streamers are concerned; it is divided into three parts of which the most notherly is broadest and strongest and the southernmost quite faint. Extreme width is 8°. On this plate tail can be traced to 20^h54^m, $-1°$, roughly 70° from the head."

"L 11 (May 15.948): Tail can be made out (northern branch) certainly to 70 Aquilae, and suspected as fas as θ Aquilae; this portion is several degrees north of the radius vector. Distance from head 80°. The northern branch passes centrally between 1 and 70 Aquilae."

The Johannesburg observers describe the head on this date as of "unusual appearance, most glow behind the nucleus" (May 15.625). On May 15.655 (43 minutes later) the nucleus was found changed and there were well marked rifts in the head (Innes). The nucleus aggain was yellow. Worssell at 15^h10^m (10^m later than the first observation by Innes) mentions a "line through the center of the nucleus." It is obvious from this description that changes occurred in the nucleus in a very short interval, but is difficult to make out anything definite. The same observers noted in the coma a star dimmed $0^m.7$.

The four Crossley plates of May 15 correspond in general to the Johannesburg descirption. The rifts between the emission fans are especially interesting. They are well visible on the north side of the nucleus and the nucleus itself is surrounded for about 180° by a narrow band which is decidedly less dense than the field of the plate outside of the coma. It appears as if non-luminous matter had been ejected from the nucleus. This would be in agreement with Worssell's observation of a (dark) line through the nucleus.

On May 16 the Johannesburg observers saw the nucleus large and yellow. "Plates L 12 and Z 7 (May 16.960) were exposed on the tail about 80° from the head, guiding on a star. Faint traces of tail on both plates. Can be traced till it merges into the Milky Way; it passes about centally across 57 Aquilae. Total length recorded 95°—100°." (Curtis.)

N. T. Bobrovnikoff, Publ. of the Lick Observ., Vol. X VII, pt. 2, p. 373, 1931.

May 16; observations see Fig. 3, No. 39

May 16. — Plate VII, No. 30. — Exposure from 15 h. 17 m. to 15 h. 29 m. Last photograph before transit. Tail 2° 5′ wide at 5° from nucleus. Inclination to meridian 263°. The greater brightness of the South edge of the envelopes and sheath is noticeable to more than 1° from the nucleus. The tail, although faint, appears to be simple in character, with no noticeable rifts or rays.

May 16. — Plate XI, No. 39. — Nucleus large and yellow.

Transvaal Observ., Circular No. 4, 11th July 1910, p. 26/29.

Le 16 Mai, à 5^h a.m., la longueur de la queue était évaluée à plus de 80° et l'éclat était estimé supérieur à la première grandeur, car une couche de brouillard s'etant interposée, la tête de la comète a continué à être visible, ainsi que Vénus, pendant que toutes les étoiles avaient disparu. Le 17 au matin, la longueur de la quetie a paru dépasseur 110°, car à 3^h la partie visible avait déja près de 90° et le lever du noyau ne devait avoir lieu qu'a 5^h55^m.

H. Morize, Astron. Nachr. Nr. 4443, S. 36, 1911.

May 17

May 17. — Nucleus rose at 5 h. 45 m. Standard Time (dawn). It looked like a mere ball of light 30″ in diamter, flashing yellow and red. At 6 h. daylitht, the nucleus was large, gut it was so near the horizon that no detail could be seen. On this and previous days the comet was invisible in a 3-inch at meridiantransit.

Transvaal Observ., Circular No. 4, 11th July 1910, p. 26.

In Dairen in der Mandschurei ($\lambda = 8^h6^m38^s$ östlich v. Gr., $\varphi = +38° \ 59'.7$) wurde der Komet Halley von Herrn K. Saotome, Astronom der k. Sternwarte Tokyo, von Mitte Mai bis Mitte Juni auf einem temporären Observatorium beobachtet. Es wird im folgenden eine Zusammenstellung der Resultate, ausgezogen aus den der Redaktion gütigst übersandten Nummern 419—451 der „Mancuria Daily News", gegeben. Die Beobachtungen wurden mit bloßem Auge angestellt.

Bürgerliches Datum	Kern		Schweif			Bemerk.
	Helligkeit	wird sichtbar	Länge	Breite	wird sichtbar	
1910						
Mai 6 morg.	2^m	2^h45^m	15°	—	—	wolkig
7 morg.	—	2 50	—	—	—	wolkig
8 morg.	2	2 40	15	—	2^h15^m	
9 morg.	2	2 40	20	—	2 45	wolkig
10 morg.	—	2 50	23	—	2 36	wolkig
11 morg.	2	2 37	25	—	2 20	
14 morg.	2	2 42		—	1 40	
16 morg.	—	4 00	57	—	2 43	wolkig
18 morg.	—	—	(96)	4°	1 48	
19 morg.	—	—	(105)	6	1 55	
20 morg.	—	—	(110)	5	—	
21 morg.	—	—	(125)	4	—	
abds.	2	7 40	30	—	—	
22 abds.	4	7 35	—	—	—	
23 abds.	3	7 45	30	—	—	
25 abds.	2	7 50	35	—	—	
26 abds.	2	7 50	40	—	—	
27 abds.	4	7 50	35	—	—	

Astron. Nachr., Nr. 4431, S. 242, 1910.

May 18/19

The transit across the Sun occurred on May 18. Ingress occurred in p.a. 279° at 15^h40^m GMT and egress in p.a. 76° at 16^h38^m (Leuschner, Report), but there was as much as 10^m difference between the various computers as to the time of the transit (see Report). The result of the observations was entirely negative. It will be discussed later in connection with the structure of the tail.

N. T. Bobrovnikoff, Publ. of the Lick Observ., Vol. X VII, pt. 2, p. 374, 1931.

Nach Crommelin (A.N. 4379) geht 1910 Mai 18.6 M.Z. Greenwich der Halleysche Komet in seinem niedersteigenden Knoten vor der Sonne vorüber. Diese Erscheinung wird vielleicht in Ostasien sichtbar sein.

J. Franz, Astr. Nachr., Nr. 4408, S. 287, 1910.

Telegramm aus Taschkent, 1910 Mai 20 nachm.
Komet Halley 18. May 21^h Taschkent durch leichte Wolken auf Sonnenprojektion bemerkbar.

Sýkora

Astron. Nachrichten, Nr. 4414, S. 369, 1910.

Der Halley'sche Komet auf der Projektion der Sonne

Während des Kometendurchganges war die Sonne in Taschkent nur bisweilen durch leichte Wolken in den letzten 10 Minuten des Durchganges sichtbar.

Die Beobachtung des Durchganges wurde mittels Projektion an einem dreizölligen Fernrohr mit 30-facher Vergrößerung ausgeführt.

Die Sonnenscheibe wurde in voller Dunkelheit projiziert, und um $20^h.95$ MZ Taschkent gelang es mir und meiner Frau, die Projektion des Halleyschen Kometen zu bemerken.

Der Komet zeigte sich in der Projektion wie ein Nebelfleck (wie Fingerzeichen auf Papier).

Man konnte zwar nur einige Minuten beobachten, aber während er Beobachtungszeit verminderte sich der Abstand des Nebelflecks von dem nordöstlichen Rande der Sonnenscheibe.

Der Austritt wurde der Wolken wegen nicht bemerkt.

Der Durchmesser der Sonnenscheibe auf der Projektion war 13 cm und der des Kometen ein wenig größer als 1 cm.

Die Größe der Bewegung während der 3 Minuten der Beobachtung betrug ungefähr $^1/_2$ cm, ihre genaue Richtung konnte man nicht bestimmen.

Auf Grund vieljähriger Beobachtungen der Sonne kann ich behaupten, daß die beste Methode der Sonnenbeobachtungen die der Projetion in voller Dunkelheit ist und daß es nur dank dieser Methode gelungen ist, den Kometen zu erkennen.

Was den Himmelszustand betrifft, so ist bekannt, daß man manchmal vor Sonnenuntergang durch schichtförmige Wolken gute Bilder bekommt. In der Tat waren bei der Beobachtung die Flecken, Fackeln und Poren auf der lichtschwachen Sonnenscheibe sehr deutlich sichtbar.

Die Figur auf der beiliegenden Tafel stellt annäherungsweise die Lage und das Aussehen des Kometen dar.

Taschkent, Sternwarte, 1910, Mai.

J. Sýkora, Astron. Nachrichten, Nr. 4431, S. 241/242, 1910.

Mitteilung von Dr. V. Cerulli, Teramo, 1910 Mai 20.

Auf der Collurania-Sternwarte, in einer Höhe von 400 m über dem Meere, machte ich am 18. Mai 1910 folgende Wahrnehmungen über den Schweif des Halleyschen Kometen.

14^h40^m M. E. Z. Der Schweif wird wahrgenommen als ein schmaler, einfacher Streifen vom Ostpunkte des Horizonts bis ε Pegasi. Der Winkel der Schweifrichtung mit dem Horizont scheint etwas kleiner als 45° zu sein.

14^h42^m. Schweif ungefähr so weißlich und ebenso hell wie die Milchstraße.

14^h47^m. Nimmt an Ausdehnung zu. Geht jetzt über ε Pegasi hinaus.

14^h53^m. Prächtiger Anblick. Der Schweif des Kometen ist sehr breit in der Nähe des Horizonts (Andromeda) und verengt sich allmählich, bis er im Azimut von α Aquilae (etwas unter diesem Stern) spitzig endet. Er gleicht einer Spindel, die mehr als 50° gegen die Milchstraße geneigt ist. Sie reicht aber nicht ganz bis an die Milchstraße. In dem Punkte des Horizonts, welcher diametral der dicksten Stelle (Aufgangsstelle) des Schweifes gegenüber steht, ist keine Spur von weißlichem Schimmer zu sehen.

14^h58^m. Eine Sternschnuppe läuft den letzten (spitzigen) Teil des Schweifes entlang. Sie verliert sich rechts vom Altair.

15^h0^m. Der Schweif scheint plötzlich in Abnahme begriffen: jetzt erstreckt er sich kaum über den Horizont hinaus, obwohl die Morgendämmerung noch nicht störend wirken kann.

15^h15^m. Die Dämmerung ist schon zu weit fortgeschritten, als daß noch etwas vom Kometenschweife wahrgenommen werden könnte.

Die rapide Verkürzung des Schweifes gegen 15^h0^m halte ich für reell und nicht für einen Effekt des Tagesanbruchs, welcher kaum geahnt werden konnte. Höchstwahrscheinlich ging in jenem Augenblicke die Schweifachse dicht an der Visierlinie vorbei. Ob die Erde aber von dem Schweife berührt wurde oder nicht, muß ich aus Mangel an einem deutlich erkennbaren Gegenschein am Horizont dahingestellt sein lassen.

Astron. Nachr., Nr. 4416, S. 403/404, 1910.

The Transit of Halley's Comet

The Transit of Halley's Comet
By *A. B. Chatwood.*

Extract from a letter to the Astronomer Royal, dated Secunderabad, Deccan, 1910 May 19: —
Transit of Halley's Comet.
Very careful visual observation at this time failed to reveal the presence of the comet, and the immergence and emergence could not be detected. Observations were made with the 8-in Cooke 133-in. focal length triplet and sinlge achromatic eyepiece 1′25-in. focal length and others. I cabled you as follows: —
Transit no trace comet seen
— and trust this negative information will be useful. Chatwood.

Halley's Comet: Transit across the Sun, 1910 May 18
By *W. E. Cooke*, M. A.

Two sets of ephemerides were available, one giving 14^h22^m G.M.T. as the time of first contact, and the other, from Popular Astronomy, April 1910, p. 240, giving 15^h50^m. The weather has been unusually cloudy for some weeks, and the first possible chance of an observation for several days occurred last night, when a hasty setting gave the approximate position of
R.A. 6^h56^m 10° and Decl. + 15° 28′ on May 21, 22^h5^m G.M.T.
(A 10″-exposure photograph was also obtained, from which the exact position will be computed.)
This approximate place agrees so closely with the Popular Astronomy ephemeris, that we may conclude that the transit across the Sun's disc commenced at 15^h50^m and finished at 16^h48^m. This was fortunate, because during the previous two hours, or from 14^h to 16^h, not a glimpse of the Sun was visible. After 16^h, however, several breaks occurred in the clouds, and clear view of the Sun were occasionally obtained. It was viewed by projection on the prepared surface of a Watson "new helioscope projection camera" attached to the 10-inch guiding telescope of the astrograph, and not only the spots but the granules were plainly visible. By giving a slow oscillating movement in R.A. to the telescope the surface markings were rendered still more clear. The results were negative, and not the slightest sign of a shadow or spot was perceptible.

Perth Observatory: 1910 May 23.

Monthly Notices of R.A.S., LXX, 8 (1910) 614.

Halley's Comet and its Spectrum as observed at Kodaikánal

By *J. Evershed* (Plate 20.)

Halley's Comet was first observed by me as morning star on April 18, at about 4^h30^m a.m. I.S.T. It then appeared as a 2nd magnitude star, with a tail about 5 degrees in length. This early appearance of the comet as a conspicuous naked-eye object had not been anticipated by us, and our preparations for phothographing the comet and its spectrum were by no means complete. We had, however, already available for direct pictures of the comet a portrait camera of 4 inches aperture and 35 inches equivalent focus attached to the North Dome equatorial, and this was immediately brought into use by Professor Michie Smith.

It seemed desirable to get photographs of the comet on both a larger and a smaller scale than that determined by the focal length of the portrait lens. For the larger scale I had available a $9^1/_4$-inch silver-on-glass parabolic mirror, and a 16-inch coelostat lent us for another purpose by the Joint Eclipse Committee of the Royal and Royal Astronomical Societies.

A teakwood skeleton tube was built and mounted on an iron base as a horizontal telescope. A small brass box holding plate $3^1/_4$ in. $\times$ $3^1/_4$in. was mounted near the open end of the telescope, and directly in the focus of the mirror. This box was very little larger than the plates, and obstructed about one-seventh of the incident light. For guiding, a refracting telescope of 3 inches aperture and 36 inches focus was attached to the reflector, the coelostat mirror being large enough to supply light to both instruments.

Guiding in declination was effected by moving the telescope in azimuth with a fine threaded screw, and in R.A. by means of an endless cord controlling the slow motion of the coelostat.

The telescope was mounted east of the coelostat, and directed towards the place of setting of the comet.

Owing to the small elongation of the comet west from the Sun, exposures much exceeding one hour were never possible, and most of the plates were not exposed longer than half an hour.

Excepting for certain irregularities in the driving of the coelostat, the cause of which was not discovered, the whole apparatus worked extremely well, and good photographs of the comet were secured on 14 days between April 21 to May 17 inclusive. (Detailed particulars regarding these photographs, and the dates and times of exposure, will be found in a forthcoming bulletin of the Kodaikànal Observatory.)

Owing to the small field of the reflector, the length of tail shown does not exceed 2 degrees in any of the plates. The most interesting plates are those obtained on April 22 and on May 13, reproduced on Plate 20, Figs. 1 and 2. On the former date a strongly marked streamer is shown extending in a wavy line along the northern edge of the tail. It is very bright near the head, bends sharply outwards at 36′ from the nucleus, then follows the general direction of the tail in a wavy line for another 25′, when there is another sudden inflection or wave. It then continues in a nearly straight line to the edge of the plates inclined some 10 degrees to the general direction of the tail. Many more or less straight streamers are shown in the main tail, but these are interrupted by a patch of light extending diagonally across the tail from the first point of inflection in the northern streamer. A comparison of this photograph with others obtained elsewhere on the same date would probably give interesting information regarding the movement of matter in the tail.

The plate of May 13 is noteworthy for the well-defined edge of the coma on the northern side, and the large number of narrow, nearly straight streamers composing the tail.

The plate exposed on May 17 was somewhat fogged by the morning twilight. The head of the comet on that day rose so late that it was found necessary to continue the exposure, even in a rapidly brightening sky. The photograph shows the coma near the head about 40′ in width, with a central condensed patch containing the nucleus about 15′ in diameter.

The comet on this date was distant from the Earth about 20 millions of miles, but on May 19, the day of transit across the Sun, the distance would be reduced to about 15 millions. If the actual dimensions of the comet remained the same, the condensed region of the comet would have covered two-thirds of the Sun's during the transit, whilst the coma would have extended considerably beyond the Sun' limb on all sides. It was therefore realized, on seeing this photograph, that the transit would be, so to speak, a total eclipse of the Sun by the comet.

A large number of small scale photographs of the comet were secured with cameras ranging from 5 to 12 inches in focal lenght, and attached to the north and south equatorials: particulars of these will be found in the publication already referred to.

The Spectrum

Two slit spectrographs were constructed specially fot the comet. One of these was attached to the South Dome equatorial, and the other was designed for use with the $9^1/_4$-inch reflecting telescope. Both these instruments were found easy to manage and adjust; but although they gave brillant images of the spectrum, I got no results of value from either of them. I attribute this want of success to the short exposure times available after the middle of May, as well as to poor conditions of seeing, the sky being generally more or less vieled by thin clouds. During the best presentation of the comet, early in May, an unexpected cyclone in the Bay of Bengal so distrubed the weather at Kodaikànal that no observations could be made.

Some fairly good spectrum photographs were obtained with a prismatic camera, specially designed for ultra-violet radiations, but these are also affected by the very moist atmospheric conditions that prevailed during May. The instrument had two 60° glass prisms of about 2 inches effective aperture, and a camera lens of $11^1/_2$ inches focus. It was attached to the South Dome equatorial, the 6-inch telescope being used for guiding.

All the plates have, in addition to the comet spectrum, a comparison spectrum of either the star Altair or Venus, from which the wave-lengths of the comet lines can be determined. To get the spectrum of the star, or planet, properly aligned with that of the comet, the following procedure was adopted. One of the cross wires in the guiding telescope being set accurately parallel to the diurnal motion, a long exposure was made on the comet, keeping the head bisected at the centre of the field the whole time. As soon as it was judged to be too light to continue exposing on the comet, a dark cloth was gently placed over the prisms, and the telescope was unclamped and turned on the star or planet, which was then brought to the upper part of the field of view, bisected by the wire, and allowed to run along the wire by the diurnal motion. The prisms were then uncovered for perhaps one minute, or until the star had nearly reached the centre of the field, when the slide was shut. The star or planet was not allowed to impress the plate on the other side of the comet, because in that position it would have interfered with the tail images. The image of Venus with the camera lens of $11^1/_2$ inches focus was small enough to give a very well-defined solar spectrum.

All the photographs taken of the spectrum are essentially similar (see fig. 3, Plate 20), and, so far as can be jugded from a preliminary examination, there appears to have been no change even in the relative intensities of the different radiations throughout the period covered by the plates, that is, from April 21 to May 16. The spectrum consists of three distinct parts, viz. —

(1) A continuous spectrum due to the nucleus in which the Fraunhofer lines are faintly shown, such as Hβ, Hγ, 4325, 4310, 4270. 4227. Hδ, H and K. These are displaced towards the violet compared with the same lines on the Venus spectrum by a measurable amount, notwithstanding the very small scale of the spectra.

A measurement of the displacement of the line 4227 on May 2 gave 108 Å, or a relative velocity of 77 km. per second, the comet approaching. On this date the comet was approaching the Earth at about 68 km. per second, but to this has to be added the velocity of recession of Venus. It is evident that the displacements is of the right order of magnitude, and that the alignment of the two spectra must have been practically perfect.

(2) A discontinuous spectrum of the gases surrounding the nucleus, but not extending into the tail.

(3) A series of monochromatic images of the tail. Of the radiations in the head region of the comet the cyanogen emssions measured at λλ 3866 and 3890 are by

far the strongest in the whole spectrum. They appear as distinct black spots in negatives which, from under-exposure, show nothing else in the spectrum whatever.

All the "head" lines which I have been able to detect on these plates are given in the following list of approximate wave-lengths. In this list a vertical line connecting two or more wave-lengths indicates a shading extending between, so that the separate lines can barely be distinguished. The micrometer thread was set at the points of maximum intensity.

Head Spectrum

	Intensity	Remarks
3866	40 }	Cyanogen.
3890	50 }	
4015	10	
4041	2 }	Probably two lines.
4050	2 }	
4197	5?	Probably two lines.
4212	10	
4256	3	
4361	? }	Swan spectrum (Fowler).
4371	8 }	
4631	?	Very faint.
4646	?	Very faint.
4682	10	Swan 4685 + CO 4679 (Fowler).
4693	5	Swan spectrum.
4723	5	Swan spectrum.
4739	5	Swan spectrum.
4745	.. ?	The Swan band seems to 4745, but this may be partly a photographic effect.

The intensities given in column 2 are exceedingly uncertain: they are intended merely to give a very general idea of the spectrum. The wave-lengths must in most cases be regarded as uncertain by two or three units. A further set of measures has yet to be made, so that this list is provisional only, and may be slightly modified. The correction for the relative motion of Venus and the Comet has been applied.

From the nature of these objective prism spectra the wavelengths of the tail lines cannot be determined with accuracy. In the following list the positions are simply read off as carefully as possible from the spectrum of Venus, no correction being applied for the relative motion, since this scarcely exceeds one Angström unit.

For comparison, I give the mean wave-lengths of the pairs of lines measured by Fowler in the low-pressure spectrum of carbon monoxide, and in the last column the sum of the intensities of each pair taken from his table.

Tail Spectrum

λ Comet	Remarks	CO (Fowler)	Intensity
3385	Strong.	3594	5
369	Faint abroad, perhaps double.	3700	6
379	Faint, narrow	3789	12
4014	Very strong, coincides approximately with line in head.	4011	14
413	Faint.	—	—
426	Strong.	4264	18
456	Strong.	4558	11
473	Faint.	4702	6

It will be seen from this comparison that all the lines measured by Fowler are represented in the comet spectrum, and the positions of the comet lines agree remarkably well with those of the CO spectrum, considering the very small scale of the comet spectra, and the difficulty in estimating the positions of the centres of the tail images.

The relative intensities also are in fair agreement, excepting the line 3585, which is much too strong in the comet. The displacement of this line towards the violet compared with CO may be due to an erroneous estimate of its position by me, but I am inclined to believe that it points to a compound origin including the CO radiation.

The stronger tail images can be traced for about 3 degrees from the head on the plate of May 1st. On May 3rd they appear forked, although the direct photographs of the tail do not show this. The distance to which these monochromatic tail images can be traced on the plates seems to depend on the exposure time and on the condition of the sky during the exposure. I have little doubt that the discontinuous emission really extends to the extreme limits of the visible tail. But a change in the relative intensities of the different images might be anticipated in the outer parts of the tail, bearing in mind the interesting experiments of Professor Fowler on the spectrum of carbon monoxide.

The entire spectrum of Halley's Comet, so far as I have been able to observe it, seems to be identical with that of Comet 1907 d (Daniel); and although it may be said that these spectra of Halley's Comet show very little; if anything, that is new, yet it is of interest to have established, I think, on a firmer basis than hitherto, the close correspondence of the spectrum as a whole with the spectra studied by Professor Fowler under varying conditions of pressure, of carbon monoxide with traces of nitrogen.

(1910) 605.

Beobachtung des Halleyschen Kometen auf dem Sonnwendstein

Die scheinbaren Dimensionen des Schweifes waren in den ersten Morgenstunden der nachbenannten Kalendertage folgende:

Mai 12 Länge 32° Schweif gerade und schmal
Mai 15 Länge 55 Schweif am Ende 5° breit
Mai 16 Länge 68 Schweif am Ende 5°—6° breit
Mai 17 Länge 83 Schweif gegen das Ende zu perspektivisch verjüngt, an der breitesten Stelle 5°, am Ende 3° breit
Mai 18 Länge 135°
Mai 19 Länge 140°

Der Kopf des Kometen konnte in diesem Zeitraume nie gesehen werden; die Bestimmung der Länge des Schweifes geschah durch Vergleichen der Ephemeridenörter des Kometen mit den Positionen der Sterne, welche am Ende des Schweifes lagen.

Die Beobachtungen von Mai 12 bis Mai 17 wurden von Herrn Dr. Rheden allein, am 18. auch von Herrn Professor Hartmann (Göttingen) und am 19. überdies noch von den Herren Dr. Palisa, König, Satori und dem Unterzeichneten angestellt.

Am 19. Mai hatte der Schweif fast genau dieselbe Lage wie am 18. Mai und schien an derselben Stelle der rotierenden Sphäre zu enden. Am 20. Mai war am östlichen Himmel trotz vollkommener Klarheit in der kritischen Gegend keine Spur eines Lichtschimmers mehr wahrzunehmen; es hatte daher der Vorübergang des Schweifes, wenigstens zum größeren Teile, schon stattgefunden.

Am Abend des 19. Mai war der Ausblick nach Westen durch Wolken und Dunst behindert, so daß vom Kometen nicht gesehen werden konnte.

J. v. Hepperger

Beobachtungen des Halleyschen Kometen zur Zeit seiner Erdnähe

Da die klimatischen Verhältnisse Göttingens sowie die Lage der Göttinger Sternwarte inmitten hell beleuchteter Straßenzüge wenig Aussicht für eine erfolgreiche Beobachtung des Halleyschen Kometen zur Zeit seiner schwierigen Sichtbarkeit tief am Morgen- und Abendhimmel bot, habe ich für diese Tage den im Semmeringgebiete gelegenen 1523 m hohen Gipfel des Sonnwendsteins als Beobachtungsstation aufgesucht. Diese Wahl ist, wie sich nun gezeigt hat, recht glücklich gewesen. Denn während der Komet in Göttingen, ebenso wie fast im ganzen übrigen Deutschland, nicht zu sehen war, ist auf dem Sonnwendstein, in erster Linie durch Herrn Dr. Rheden, der im Auftrage der Wiener Akademie der Wissenschaften dort in einem kleinen Observatorium Untersuchungen über die Beschaffenheit der Sternbilder ausführt, eine wertvolle zusammenhängende Beobachtungsreihe erhalten worden. Ich selbst kam, da ich den Kometen zuvor an anderen tiefer gelegenen Orten zu Gesicht zu bekommen versuchte, erst am 17. Mai auf dem Gipfel des Sonnwendstein an. Im folgenden teile ich die von mir ausgeführten Beobachtungen mit.

Am Morgen des 18. Mai (bürgerl.) war der Himmel ziemlich stark mit dünnem Gewölk überzogen. Durch eine größere Lücke konnte jedoch das am höchsten liegende Ende des Kometenschweifes gut erkannt werden. Es hatte die Helligkeit der schwächeren Teile der Milchstraße und verschwand 2^h50^m M.E.Z. in der Morgendämmerung. Eine recht eigentümliche Erscheinung bot der Schweif dadurch dar, daß er sich nach dem Ende hin nicht, wie man zunächst hätte erwarten sollen, ausbreitete, sondern vielmehr zu einer Spitze, die nahe bei ϑ Aquilae lag, verjüngte. Aus dieser Beobachtung folgt eine Schweiflänge von 100°.

In der Nacht vom 18. zum 19. Mai (bürgerl.) hielten wir — es beteiligten sich an den Beobachtungen außer Dr. Rheden und mir noch vier Wiener Astronomen — den Himmel dauernd unter Beobachtung. Der Himmel war, wenn auch nicht völlig rein, so doch für die Beobachtungen ziemlich günstig. Zunächst wurden außer sechs sporadischen Sternschnuppen während der ganzen Nacht keine auffälligen Erscheinungen, wie Aufhellung des Himmels oder einzelner Teile desselben, bemerkt. Zu unserer Überraschung zeigte sich jedoch, als die Erhellung des Himmels durch den Mond aufhörte (gegen 2^h früh), der Schweif des Kometen in beinahe unveränderter Lage und Helligkeit wie am vorhergehenden Tage.

In der beigefügten Skizze habe ich die Begrenzungslinien des Schweifendes, wie er sich am Morgen des 19. Mai (bürgFrl.) um 2^h30^m dem unbewaffneten Auge darstellte, eingetragen. Die Zeichnung beruht auf folgenden von mir genommenen Alignements. Die nördliche Begrenzung verlief mitten zwischen α Andromedae und γ Pegasi hindurch, nördlich von α Pegasi, südlich von ε Pegasi nach ϑ Aquilae hin. Die südliche Grenzlinie ging durch γ Pegasi, durch einen Punkt, der auf der Verbindungslinie von β nach α Pegasi um ein Drittel der Entfernung dieser beiden Sterne voneinander jenseits von α Pegasi lag, und verlief ebenfalls nach ϑ Aquilae hin. Es folgt hieraus eine Schweiflänge von 112°, wovon etwa die Hälfte in der Figur dargestellt ist; die tiefer liegenden Teile und der Kopf wurden von einer Dunstschicht verdeckt. Die Mittellinie des beobachteten Schweifstückes liegt mit dem Kometenkopfe nicht auf einem größten Kreise. Der Schweif erschien also nicht völlig gerade, sondern an seinem Ende nach Süden gebogen. Um 2^h40^m verschwand der Schweif in der Morgendämmerung.

Am Abend des 19. Mai konnte wegen Bewölkung des westlichen Horizontes keine Beobachtung des Kometen erhalten werden.

Am Morgen des 20. Mai (bürgerl.) stellten wir bei recht reinem Himmel fest, daß von dem Schweife keine Spur mehr an der früheren Stelle aufzufinden war. Der Vorübergang des Schweifes an der Erde war somit nunmehr erfolgt.

Am Abend des 20. Mai fand ich den Kopf des Kometen mit einem kleinen

Kometensucher ohne Kreise um $8^h 9^m$ auf. Zehn Minuten später konnte man den Kometen auch mit bloßem Auge in der hellen Dämmerung gut erkennen. Der Eindruck für das bloße Auge war der eines tief stehenden Sternes erster bis zweiter Größe. Im Sucher erkannte man den gelb gefärbten Kern von etwa 1' Durchmesser und den hellsten Teil des auf die Sonne zu gerichteten sichelförmigen Kopfes. Erst später wurden noch schwache Spuren eines kurzen Schweifes bemerkt.

J. Hartmann. Astron. Nachrichten, Nr. 4414, S. 369—373, 1910.

Während des Durchganges des Halleyschen Kometen über die Sonnenscheibe sind von Herrn Prof. Schwarzschild am großen Refraktor des Potsdamer Observatoriums mit dem Spektrographen Nr. 1 zwölf Aufnahmen des Sonnenspektrums gemacht worden, zwecks Entscheidung der Frage, ob das Sonnenlicht beim Durchgange durch den Kometen eine merkliche selektive Absorption erleidet. Diese zwölf Aufnahmen zerfallen in zwei Serien zu je 6 Aufnahmen, von denen die erste zu einer Zeit gemacht wurde, als sich der Komet, gemäß den von Prof. Kobold (A.N. 4413) angegebenen Durchgangszeiten in der Nähe des Mittelpunktes der Sonne befand, während sich die zweite Serie auf das Ende des Durchganges bezieht, als der Komet in unmittelbarer Nähe des Sonnenrandes stand. Da die der Ephemeride entnommenen Deklinationen des Kometen genauer waren als die Rektaszensionen, wurde der Spalt des Spektrographen in der Richtung der stündlichen Bewegung eingestellt, und dann in rascher Aufeinanderfolge eine Reihe von Aufnahmen bei konstanter Rektaszension und in etwas verschiedenen Deklinationen gemacht. Die Länge des Spaltes im Bogenmaß betrug ungefähr $3^1/_2{}'$, so daß ein ziemlich großes Gebiet der Sonne überdeckt wurde. Die Expositionsdauer betrug etwa $^1/_4{}^s$. Die 6 Aufnahmen der 1. Serie wurden in der Zeit von $4^h 55^m$ bis $5^h 10^m$ (mitteleuropäische Zeit), die der zweiten in der Zeit von $5^h 28^m$ bis $5^h 30^m$ gemacht.

Um die Genauigkeit der Einstellung auf den unsichtbaren Kometen zu kontrollieren, sind aus den Elementen der Bahn (Lick Observatory Bulletin 179) zwei Kometenörter, vor und nach dem Durchgang, berechnet worden, für welche Beobachtungen vorlagen. (Es ist nicht möglich gewesen, aus den Ephemeriden von Dr. Smart, A.N. 4379, und von Fischer-Petersen, A.N. 4413, durch Interpolation brauchbare Werte zu erhalten).

Die Korrektionen der berechneten Orter ergaben sich zu:

	$\Delta\alpha$	$\Delta\delta$	Beobachtung von:
Mai 15.9960 Gr.M.Z.	$-35\overset{s}{.}1$	$-1'\ 3''$	Aitken, Mt. Hamilton (A.N. 4413)
Mai 20.3015 Gr.M.Z.	$-55{,}0$	$+2\ 28$	Battermann, Königsberg (A.N. 4414)

Durch Interpolation wurden für die Zeit des Durchganges die Korrektionen erhalten: $-47^s\!.6 + 0' 59''$. Die damit berechneten Positionen des Kometen zur Zeit der Beobachtung, sowie die jeweilige Lage des Spaltmittelpunktes des Spektrographen sind in der nachstehenden Tabelle vereinigt.

	M.Z.G.	α ☄	α_{Spalt}	δ ☄	δ_{Spalt}
Mai 18	15^h55^m	$3^h39^m37^s$)		+19°36′.8)	
	16 10	40 15)	$3^h40^m0^s$	37.3)	19°38′—19°31′
	16 28	3 41 0)		+19 37.9)	
	16 30	41 5			
	16 30	41 5)	3 41 15	38.0)	19 38 —19 32

Daraus ergibt sich, daß der Spalt sehr nahe am Kopfe des Kometen gestanden hat und ihn auch wahrscheinlich mehrmals berührt hat. Die Spektralaufnahmen, auf welchen das Sonnenspektrum vom ultravioletten Teile bis zu den D-Linien sichtbar ist (AGFA Chromo-Platten), wurden mit Hilfe des neuen Toepferschen Meßapparates untersucht. Aus drei Fraunhoferschen Linien (λ 5890.1, 5167.5 und 4005.4) wurden eine Hartmannsche Dispersionsformel abgeleitet, mit deren Hilfe im Spektrum alle Stellen aufgesucht werden konnten, an denen vom Kometen herrührende Absorptionslinien zu erwarten waren, jedoch ohne daß es gelungen wäre, ihre Anwesenheit zu konstatieren.

Besondere Aufmerksamkeit erhielten die folgenden Stellen, an denen nach Angabe des Lowell Observatory Bulletin Nr. 47, im Spektrum des Halleyschen Kometen helle Emissionslinien auftreten:

$$\lambda = 3882, 4214, 4737, 5165, 5635$$

und auch die Wellenlängen der von Dr. Rosenberg (Astroph. Journ. vol. 30, p. 271) im Spektrum des Kometen Morehouse beobachteten Strahlungen

$$\lambda = 3804, 3898, 4035, 4277, 4563, 4682, 5100, 5600.$$

An keiner dieser Stellen ist eine Abweichung vom normalen Sonnenspektrum zu sehen, so daß auch die selektive Absorption des Lichtes durch den Kometenkörper nur außerordentlich schwach sein kann.

v. d. Pahlen, Astr. Nachr., 4425, S. 138 bis 140, 1910.

Beobachtungen betreffend den Durchgang der Erde durch den Schweif des Halleyschen Kometen

The following observations were made at the Khedivial Observatory, Helwan, and may help to elucidate the question as to wheter the earth passed through the comet's tail, and if so when.

1910 May 18 13^h Gr. M.T. Tail was seen to stretch as far as α Equulei. It was about 8° broad where it involved γ Pegasi, and tapered off to a breadth of 2° at α Equulei.

May 19. No sign of tail in the west throughout the evening. At 13^h Gr. M.T. tail was traced past θ Aquilae where it was lost in the Milky Way. It could not be traced on the other side of the Milky Way. It again had the tapering form of the previous night, but was as much as 15° broad where it involved α Pegasi.

May 20. Again no tail could be seen in the evening. The head was visible to the naked eye for ten minutes only. At 14^h Gr. M.T. the tail could again be traced until it was lost in the Milky Way. To the west of the Milky Way there was a cloud which prevented the possibility of tracing it further. It was about 10° broad in Pegasus.

May 21 6^h Gr. M.T. The tail was visible to a length of 20°. At 14^h there was no sign of the tail in the east or near the Milky Way during the few minutes between the setting of the moon and the brightening of the dawn.

The narrowness (8°) of the tail on May 18, only a few hours before the predicted time of conjunction, at once suggested that the tail was not straight, but bent back in the comet's orbit. The increased breadth on the next morning seems to show that the tail was then nearer to us. The decreased breadth on the 20th may be accounted for by the fact that the part of the tail in Pegasus was no longer that nearest to us. Unfortunately none of the above nights were quite clear, and the tail could not be observed below Pegasus where on the 18th and 19th it should have been becoming apparently narrower again.

It therefore seems probable that the tail did not begin to sweep past the earth until May 20 12 Gr.M.T. at the very earliest. At this time the earth was nearly four miles south of the plane of the comet's orbit. This means that the tail passed well to the north of the earth, as the observations of the tail throughout May showed it to have nothing like the breadth necessary to envelope the earth when at this distance from the plane of the comet's orbit. Observations of the length of the tail during the week before conjunction made it consistently over 20 millionen miles, so that the earth would have been involved if the tail had been straight or nearly so.

The sun's disc was watched from May 18 15^h15^m to 16^h15^m Gr.M.T. with a four-inch Cooke equatorial, but no sign of the comet's transit was detected.

Khedivial Observatory Helwan 1910 May 29.

K. Knox Shaw

Der Kern des Halleyschen Kometen war gestern auf der Sonnenscheibe nicht zu finden. Allerdings stand mir ein größerer Refraktor nicht zur Verfügung. Die erdmagnetischen Instrumente blieben ruhig; in der Telegraphenleitung wurden Erdströme nicht bemerkt. Meteorologisch nichts auffälliges bemerkt, außer einem Gewitter in der Nacht vom 18. auf den 19., morgens 4^h (Küstenzeit). Auch luftelektrische Messungen ergaben nichts besonderes.

Tsingtau, 1910 Mai 20.

Dr. Meyermann

Astron. Nachr., Nr. 4418, S. 31/32, 1910.

Observations of the Tail of Halley's Comet before and after the day of Transit
By *J. Evershed*

During the first two weeks in May, Halley's Comet presented a fine spectacle on clear mornings, and had already attracted a considerable amount of public attention and interest. On May 12 I measured the tail directly with a sector used for determining altitudes and found it to be 60 degrees in length.

On the mornings of May 17 and 18, which were quite clear at Kodaikánal, the tail was a magnificient object between 3 a.m. and 5 a,m, Passing over the star θ Aquilae, it could be traced as far as the Milky Way, or more than 100 degrees from the head. The portion of the tail nearest to the Earth, and 90 degrees from the Sun, did not appear any wider on the 18th on the previous day, although its actual distance must have decreased greatly during the interval.

On the 19th the sky was much obscured by hazy cloud, through which, nevertheless, the tail could be plainly seen. Its width was estimated to be $6^1/_2$ degrees at an altitude of about 20°.

No observation was made on the morning of the 20th on account of clouds, but on the evening of that day the head of the comet was seen after sunset, with a short tail (evidently much foreshortened) directed in the normal way opposite to the Sun.

On the 21st, when at sea off Colombo, I again saw the tail in the east at 4.30 a.m. It appeared no broader than on the 18th, but I think it was fainter. Passing centrally through the square of Pegasus, which was nearly filled with the faint light, the tail could be traced, as before, right up the Milky Way. The star ε Pegasi was nearly in the centre of the band of light, and the star θ Aquilae near its southern edge.

This was the last observation I was able to make before dawn. It seems very remarkable that the tail should have remained visible in the morning sky, as a narrow band of light, nearly two days after the head of the comet had passed to the other side of the Sun. A part of the tail on the 18th must have been rapidly approaching the Earth, and it ought to have widened out and disappeared as it enveloped the Earth.

At 4.30 a.m. on the 21st the head of the comet had passed far to the east of the Sun, and presumably was developing the tail seen the previous evening, directed away from the Sun towards the east. Yet the greater part of the tail remained on the west side of the Sun, apparently still at a great distance from the Earth, judging by its small width. If continuous with the head, it must have passed near the Sun's position, and might have been expected to appear as an abnormal tail on the sunward side of the comet.

No trace of a tail directed towards the Sun was seen by me on the evenings of May 20 or 21, although specially looked for, but the sky was very bright with the evening twilligh and with moonlight.

The extraordinary persistence of the light in the eastern sky may perhaps be explained by supposing the tail to be strongly curved, and very broad in the direction of the comet's motion, although narrow and straight in the direction at right angles to this. If so, the passage of the Earth through the tail, if it occurred at all, must have been delayed one or two days, and probably occupied more time than a single day. There is some doubt whether the tail did actually touch the Earth, for observations of its position in the sky on May 11 and 15 show that its axis was inclined very considerably northward from the direction of the radius vector of the comet.

Monthly Notices of R.A.S. L XX, 8 (1910) 610.

The Transit of Halley's Comet across the Sun

By *J. Evershed*

The following programme was arranged at the Kadaikánal Observatory for the observation of the transit.

(1) A careful scrutiny of the Sun's disk by direct observation with high powers, using the 6-inch Lerebour et Secretan refractor.

(2) A series if direct photographs to be taken with varying exposures and slow plates, on a scale of 8inches to the Sun's diameter.

(3) A series of monochromatic photographs to be taken with the spectroheliograph, the camera slit being set on the head of the cyanogen fluting at λ 3883.

(4) A series of monochromatic photographs to be taken with the camera slit set on the line K, for comparison with the others.

(5) At the time of mid-transit some photographs were to be taken with the

large grating spectrograph of the cyanogen bands at 3871 and 3883, observing at the same time the region of the blue "Swan" band.

Had the sky been clear, it was intended to begin the cyanogen photographs before ingress and to continue after egress, in the hope that with very long exposures the comet might appear as a bright object outside the Sun's limb. During the time of transit, exposures of about 30 seconds total duration would have sufficed to give good images of the disk, and the comet was expected to appear on these as a dark absorption marking, as is the case with many of the solar prominences.

I had great hopes of the success of these monochromatic plates, because of the very great intensity of the cyanogen radiations in the comet's head as revealed by the spectrum plates.

The time of ingress as communicated to us beforehand was 7^h51^m .. Indian standard time, and egress one hour later. Although clear skies are the rule in the early mornings at Kodaikánal, and had prevailed since January, the morning of May 19 was cloudy. At 8 a.m. there was a momentary gleam of sun, which enabled us to adjust instruments. At 8^h30^m the clouds hat thickened, and all hope of obtaining successful photographs was abandoned. But at about this time a cable message came through from Kiel which stated that ingress would not take place until 8^h59^m. This information, so oppurtunely received, saved the situation, as there occurred several moderately clear intervals between 9^h and 10^h, when it was possible to partly carry out the programme as regards items 1, 2, and 3. At no time was the Sun entirely free from cloud, and the intervals of moderate clearness were few. As a compensation, the definition, which is usually very poor after 9 a.m., remained good. No long-exposure plates before ingress or after egress were attempted on account of the thick clouds surrounding the Sun.

Results. — No trace of the comet can be seen of any of the direct photographs which were obtained, and Professor Michie Smith could see nothing whatever of an unusual nature on the Sun's disk with the 6-inch telescope, althoughhe states that the definition during the partially clear intervals was excellent.

Eight monochromatic photographs of the disk in the cyanogen radiation 3883 were obtained at the following times: —

No. of Plate	I.S.T. h m	Remarks
794	8 59	Fair.
796	9 32	Fair.
797	9 35	Poor, much cloud.
798	9 38	Fair, good definition of details.
799	9 56	Fair, good definition of details.
800	9 59	Poor, much cloud.
801	10 02	Very poor and under-exposed.
802	10 19	Good.

All the plates are much steaked with cloud, but Nos. 796, 798, 799, and 802 are otherwise good, and show much minute detail all over the disk. If the times of ingress and egress transmitted to us from Kiel are not greatly in error, the comet should appear on the three plates Nos. 796, 798 and 799. A preliminary examination gives a negative result; and if the comet nucleus is shown at all, it must be too small or faint to be readily distinguished from the innumerable small markings, both dark and light, shown in all parts of the disk. On the other hand, the cyanogen part of the comet may be too large and diffuse to appear as a definite dark marking. The head of the comet emitting the cyanogen rays must have been larger than the Sun's disk, and it is probabble that a more or less uniform dimming of the entire disk would escape notice, especially under such cloudy conditions.

Plate No. 802 is perhaps the best of the series, but it was exposed about 20 minutes after egress: this photograph does show a small dark spot, unlike a defect in the film, at 7!6 north from the centre, in position angle 16° towards the east. The spot is about 7″ in diameter, and is quite opaque, but has ill-defined edges. A cloudy interval of 18 minutes preceded No. 802, and no more exposures were possible for over an hour afterwards. I had no opportunity, therefore, to ascertain whether this marking was moving at the same rate as the comet: it is certainly not shown on the earlier plates, and no trace of it can be seen on a direct photograph taken about two minutes earlier.

Monthly Notices of R.A.S. L XX, 8 (1910) 612.

Transit of Halley's Comet, 1910 May 19. Observations made at Sydney Observatory

By *W. E. Raymond*

The $11^1/_2$-inch refrator of the Sydney Observatory was used on May 19, 1910, between 12 p.m. and 2 p.m., in the endeavour to trace Halley's Comet as it crossed the Sund's disc (time of crossing 12.22 to 1.22 p.m.).

Messrs. Walter F. Gale, of Newcastle (who had just returned from the Eclipse expedition to Brune Island, Tasmania), R. P. Sellous, B. A. Lands Department, Sydney, and the writer spent over an hour watching the Sun's disc, but not a trace of the comet could be found. Several patches of sun-spots were observed, and some of them had to be watched very minutly before any definite announcement could be made as to the possibility of their being the nucleus of the comet. Clouds interfered slightly with the observations at 1.12 p.m., but cleared again before the time announced for the termination of the transit.

Mr. W. J. McDonnell, of Mosmar, Sydney, telephoned to state that he had made a careful search of the Sun with his refractor ($4^3/_4$-inch), but could detect nothing in the way of the comet.

Sydney: 1910 June 9.

Monthly Notices of R.A.S. L XX, 8 (1910) 653.

Observations of Halley's Comet made on board s.s. "Romanic"

By Captain *H. P. David* and *H. N. M'Gill*, Fouth Officier. (Communicated by Capt. *Campbell Hepworth*.)

May 18. 1910. Position 40° 51′ N., 43° 34′ W.

At 2^h45^m a.m. observed luminous tail of Halley's Comet, rising from a point 5° above horizon; on true bearing of N. 80° E., and extending upwards into sky at an angle of 35°.

Tail appeared to preserve an uniform width of 4°, and passed through lower corner of square of Pegasus, the star γ Pegasi appearing plainly through it.

It extended nearly to α Aquilae, where it became invisible to the naked eye.

Cumulus clouds obscured the horizon to an altitude of 6°, which may account for not seeing the comet itself.

The planet Venus rose at 3^h30^m a.m., ship's time, appearing exceptionally brilliant, though daylight was rapidly coming in.

At 3^h45^m a.m., when daylight had fully broken, the tail became invisible. It was seen at its best at 3^h15^m a.m., when the Moon had set.

Ship's time 8^h24^m p.m. = 9,10 p.m. G.M.T.

Observed Halley's Comet bearing N. 85° W. (true), altitude 18°.

Angular distance with α Canis Minoris 9° 46′.

Angular distance with β Geminorum 14° 45′.

Comet, as compared with bright stars near, appeared about the same size as Jupiter, but with little luminosity, and of a distinctly yellowish tinge.

There was little or no to be seen, probably owing to the Moon being at the full and shining brightly.

Meteorological Office, London, S.W.: 1910 June 6.

Monthly Notices of R.A.S. L XX, 8 (1910) 618.

Photographs of Halley's Comet made at the Córdoba Observatory, Argentine Republic

(Extract from a letter from Prof. D. C. Perrine.)

I am sending to the Royal Astronomical Society some copies (enlarged nearly two times) on glass, of negative of Halley's comet made here.

We secured a long series of good photographs of the comet during its greatest brightness, of which the following is a summary: —

Position plates with the astrographic telescope:

April 18—30	9 nights.
May	22 nights
June	23 nights
July	14 nights
August	3 nights.

Long and short exposures for structure were obtained on practically the same nights to and including the first week in July, after which the comet was too faint to justity other than position plates.

An equally extensive series was obtained with the portrait lens of 5-inch aperture.

Extra-focal observations were also obtained on 40 nights for determinations of brightness.

Observations were secured with two small objective prism spectrographs on 40 nights between May 21 and July 28. One of the spectrograms recorded the spectrum of the dislocation in the tail observed on June 6. A casual examination shows no difference of constitution between this disturbance and the normal tail of the night before.

Photographs of this disturbance were obtained on June 6, 7 and 8, which will give good values of the velocities of recession from the head.

A small camera was improvides to record the great lengths of tail visible before the transit in May.

Micrometer observations were began on 1909 November 30, and obtained on nights between that date and 1910 February 3, inclusive, when the comet was getting too low in the west to allow of as accurate observations as could be made in the northern hemisphere. A few observations were obtained after the comet became visible in the morning sky in April, and were resumed again on August 6. Positions were obtained on 10 nights from August 6 to 22 inclusive. After the latter date it was too low in the west for satisfactory observations. The comet was seen however, on August 25.

It is clear that the bright streamer of the tail was not encountered by the Earth, but passed to the north of us. There was, however, a wide, faint streamer to the south which was observed here on the mornings of May 19 and 21 (civil time),

(May 20 was cloudy), which, if it persisted, should have encountered the Earth. On the morning of May 18 (civil time) a mistiness was observed to the east, which may possibly have been a faint tail streamer near us. If it was in reality a streamer, it was probably nearly in the direction of the radius-vector, and the Earth would have passed through it at or near the time of transit.

It is certain that the main tails lagged considerably behind the radius-vector.

I notice the suggestion of Innes that the Earth may have the power of repelling cometary tail-matter in a similar way to the Sun. It seems to me doubtful if this is the case. If this force is in reality light pressure, as seems very probably, the small intensity of the sunlight reflected by the Earth could hardly overbalance its gravitational force. If the force of repulsion is electric or of some other nature, then the case is otherwise.

The comet was a magnificent spectacle, especially during the week before the transit. That great searchlight beam, extending away from the Sun for 120° to 130°, was a sight which cannot be forgotten.

Córdoba: 1910 September 6.

Monthly Notices of R.A.S LXXI, 1 (1911) 102.

Mitteilungen vom Halleyschen Kometen

Über Wahrnehmungen in der Nacht vom 18. Mai und über weitere möglicherweise mit dem Durchgange der Erde durch den Schweif des Halleyschen Kometen zusammenhängende Erscheinungen liegen folgende Mitteilungen vor:

Von Prof. E. Hartwig, Bamberg, 1910 Mai 19.

„In der vergangenen Nacht sah ich nichts vom Schweif, wenn nicht ein durch α Pegasi um $14^1/_2{}^h$ M.Z. Bamberg gehender Streifen, der nach unten keine Abgrenzung zeigte, vielleicht doch ihm zugehörte. Die Sonne durchmusterte ich vom Aufgange an bis gegen 18^h M.Z. Bamb., ohne außer der schönen Fleckengruppe in der Mitte und einem runden Fleck am Rande, die ich gestern am Heliograph photographisch aufgenommen habe, auch nur ein Fleckchen entdecken zu können."

Von Prof. E. Strömgren, Kopenhagen, 1910 Mai 19.

„Während der ganzen Nacht (Mai 18/19) war vom Kometen nichts zu verspüren. Zur Zeit des Durchgangs durch die Sonnenscheibe beobachteten Obervator Pechüle, Assistent Nörlund und der Unterzeichnete durch Projektion mit Hilfe des kleinen Fraunhofer (Öffnung 95 mm). Mehrere Sonnenflecken, aber vom Kometen nichts zu sehen."

Von Prof. M. Wolf, Königstuhl-Sternw., 1910 Mai 20.

„Ein seit der Nacht vom 17. zum 18. Mai sich entwickelnder Cirrusschleier, der bis zum Nachmittag des 19. immer mehr an Fülle und Formen zunahm, und der Mond verhinderten jede feinere Beobachtung in der Nacht vom 18. zum 19. Mai. Es war aber auch kaum zu erwarten, daß die Erde den dickeren Teil des Kometenschweifes vor dem Vormittag des 19. berührte.

Von dem Vorübergang des Kometenkopfes vor der Sonne konnten wir, obgleich die Sonne recht gut zu beobachten war, „mit 6 Fernrohren" nichts sehen.

Dagegen scheinen die optischen Vorgänge in der Atmosphäre am 19. Mai dafür zu sprechen, daß wir wenigstens einen Teil der photographisch angedeuteten ausgedehnten Wolken des Schweifes passiert haben.

Am späten Nachmittag wurde zuerst der Bishopsche Ring um die Sonne sichtbar. Dann entwickelte sich, nachdem nur relativ geringe Dämmerungsvorgänge in der letzten Zeit beobachtet waren, am Abend des 19. eine Dämmerung von ganz ungeahnter Intensität, Ausdehnung und Dauer. Es konnten drei aufeinander folgende Purpurlichter beobachtet werden, und alle jene Erscheinungen wie um den 1. Juli 1908 und früher nach den Ausbrüchen des Krakatoa und des Mont Pelée traten in verstärktem Maße ein.

Um den Mond zeigte sich der Bishopsche Ring in einer Intensität, wie ich ihn noch nie gesehen hatte. Ich bestimmte den äußeren Radius auf 28°0 zur Zeit der Kulmination des Mondes (in 37° Höhe), was einer Größe der Teilchen von $1^1/_2$ Tausendstel Millimeter entspricht.

Die Cirrusbewölkung nimmt jetzt ständig ab, aber in den hohen Schichten bleibt eine schwache, verfilzte Granulation, die den Bishopschen Ring um die Sonne heute mit ziemlicher Intensität zu sehen gestattet.

Sternschnuppen und nordlichtartige Erscheinungen sind von uns nicht beobachtet worden, letzteres hätte der Mond auch zweifellos unmöglich gemacht.

Das Wolkenbild in den letzten Tagen war äußerst interessant. Cirren aller Arten und mit allen möglichen Strahlungspunkten zeigten sich in- und übereinander. Ganz eigenartig waren die bei schiefer Beleuchtung sichtbaren dünnen, rauchstreifenartigen Cirren, die, von S 20° E nach N 20° W gerichtet, alle anderen Systeme unbehindert durchdrangen und die Form der strahlenförmigen Nordlichter nachahmten."

Von der Sternwarte Upsala, 1910 Mai 19.

Während der Passage des Halleyschen Kometen vor der Sonne haben wir (von 16^h bis 18^h M.E.Z. Mai 18) die Sonnenscheibe in dem mit einem Polarisationshelioskop versehenen großen Refraktor der hiesigen Sternwarte (Öffnung 36 cm) genau durchmustert. Der Himmel war wolkenfrei und die Luft gut. Die angewandte Vergrößerung war meistens eine 180-fache. Vom Kometen war nichts zu sehen.

Östen Bergstrand, H. v. Zeipel.

Von Herrn W. Krebs, Großlottbek, 1910 Mai 20.

„Polarisations-Beobachtungen mit einem Polariskop Savart ließen eine vorübergehende, aber kräftige Einengung des zenitalen Feldes positiver Polarisation erkennen, die Mai 18 zwischen 15^h10^m und 15^h42^m M.Z. Greenw. einsetzte. Innerhalb dieser Zeit verkürzte sich dessen Diagonale im Vertikal der Sonne um 6°, gegenüber dem ungefähr normalen Verhalten des Jahres 1889 um 11°. Bis 18^h40^m M.Z.Gr. nahm die Verkürzung in drei Terminen fast stetig bis auf 15° zu. Eine Abendbeobachtung. Mai 19 um 7^h25^m, ergab wieder annähernd normale Polarisation, ebenso eine Morgen- und drei Abendbeobachtungen von Mai 19/20.

Die Bewölkung am Morgen des 19. Mai (bürg.) wurde von glänzenden Cirrusschleiern gestellt, die um 14^h57^m 0.2, um 15^h10^m 0.3, um 15^h42^m und 16^h11^m 0.5, um 18^h34^m und 18^h40^m 0.6 des Himmels bedeckten. Sie zogen aus SSW, während der Wind an der Erdoberfläche mit Beaufort 6 aus NE wehte. Mit dieser cirrösen Verschleierung hing das Auftreten einer Sonnensäule von 5° Höhe über der aufgehenden Sonne und einer kärftig irisierenden Nebensonne, die nördlich der Sonne, von 16^h7^m bis 16^h30^m beobachtet wurde, zusammen. Die Cirrusschleier waren zu Polarbanden angeordnet, die von SE nach NW strahlten. Innerhalb dieser Richtung ließen sie zarte Locken und Kurven erkennen, aber keine besonders ausgeprägte Streifung.

Gleichzeitig mit der Nebensonne trat eine Aureole um die Sonne auf, von 5° Radius und in eigenartig weißflimmerndem Lichte.“

Von Prof. M. P. Rudzki, 1910 Mai 21.

„Hier in Krakau haben wir während des Durchganges der Erde durch den Schweif des Halleyschen Kometen außer einer magnetischen Störung nichts beobachtet. Übrigens war der Himmel seit etwa 3 Uhr nachts wolkig.

In Ermanglung eines Variationsapparates wurde im Garten ein magnetischer Theodolit aufgestellt und die Deklination von Mittag Mai 18 bis Mittag Mai 19 vom Assistenten der Sternwarte Herrn J. Ryzner stündlich, von 3^h bis 6^h früh jede Viertelstunde abgelesen.

Zwischen 13^h und 14^h M.E.Z. oder noch früher begann eine Störung, die, wie es scheint, erst gegen 22^h sich vollständig beruhigte. Es war nämlich

Mai 18	westl. Dekl.	Mai 18	westl. Dekade
$12^h\,0^m$	5°22′10″	16^h15^m	5°24′30″
13 0	5 23 10	16 45	5 23 10
14 0	5 26 10	17 30	5 25 20
15 0	5 17 40	17 45	5 25 20
15 15	5 17 40	20 0	5 22 0

Nachher Zunahme bis 24^h. Zwischen 15^h0^m und 15^h15^m muß ein Minimum, zwischen 17^h30^m und 17^h45^m ein Maximum liegen.

In diesen Angaben sind die Kollimation des Magneten und die Torsion des Fadens berücksichtigt. Die Zahlen wurden auf 10″ abgerundet, da die Kreiseinteilung nur bis auf 10″ genau abzulesen gestattet."

Von Prof. J. Franz, Breslau, 1910 Mai 20.

„1910 Mai 19 12^h15^m M.E.Z. beobachteten L. Mewes (zweiter Assistent der Sternwarte) und F. Pavel (stud. astr.) zwei helle, am Rande verwaschene Bogen am Nordhimmel. Die Bogen entsprangen vereinigt in der Mitte der Verbindungslinie α Aurigae — α Persei und zogen senkrecht zu ihr nach oben. Der obere Bogen war heller als der untere, ihre Verlängerungen verliefen allmählich in den dunklen Himmelsgrund, der unter Bogen war fast bis $\varkappa$ und ι Ursae maj. zu verfolgen. Nach dem oberen Ende zu entfernten sich beide Bogen immer weiter voneinander, die größte Breite der ganzen Erscheinung mag 3°—4° betragen haben. Der Radius der durchaus kreisförmigen Bogen wurde auf 45° geschätzt.

Die Erscheinung bewegte sich dann weiter, derart, daß der Anfangspunkt sich immer mehr nach α Persei hin verschob, die Bogen sich grader streckten, gegen den Zenit emporstiegen und sich bedeutend verlängerten.

Es wurden folgende Aufzeichnungen gemacht, wobei der obere Bogen mit I, der untere mit II bezeichnet ist:

12^h45^m. I geht diagonal durch den großen Wagen von β nach δ und ist darüber hinaus sichtbar, II geht hart unter β Ursae maj. vorbei.

12^h51^m, I geht zwischen ε und δ Ursae maj. hindurch. Beide Bogen beginnen jetzt in α Persei. Eine Einknickung der Bogen nach Westen, die 12^h40^m nahe dem Anfang bemerkbar war, ist wieder verschwunden. Die Trennungsstelle beider Bogen wandert immer mehr nach oben.

12^h52^m. II geht durch β und δ Ursae maj.

12^h54^m. I geht zwischen η und ζ Ursae maj. hindurch und ist auf den Mond gerichtet. II zwischen ε und δ.

12^h57^m, I geht der Länge nach durch den kleinen Wagen, ist weit über Arktur hinaus zu verfolgen.

12^h59^m. II geht zwischen η und ζ Ursae maj. hindurch, ε Cassiopeiae tritt in I ein.

13^h0^m. I ist auf Spica zu gerichtet.

13^h4^m. Arktur zwischen I und II.

13^h9^m. II durch Gemma. Die ganze Erscheinung ähnelt in ihrem Aussehen sehr der Milchstraße, ist aber bedeutend heller. Sie beginnt zu verbleichen. Ganze Breite im Zenit etwa 10°, Länge 130°.

13^h10^m. Beginn in der Cassiopeia, Trennungsstelle etwa im Zenit. Sehr verwaschen.

13^h13^m. Nur noch vom Zenit bis zur Cassiopeia zu sehen.

13^h15^m. Um Wega am deutlichsten, sonst nur noch ein schwacher Schein, wenig heller als die Milchstraße.

13^h20^m. Fast erloschen.

Auch in ihrer größten Ausdehnung um 13^h8^m bildete die Erscheinung einen vollkommen glatten Bogen ohne jede Knickung, der wohl einem größten Kreis am Himmel entsprach. Trotz des Mondlichtes war er etwa bis Spica zu verfolgen.

Der Himmel war durchaus klar, nur im Westen stand eine langgestreckte Cirre, die aber bedeutend schärfer begrenzt erschien als die oben beschriebenen Bogen. Es scheint ausgeschlossen, daß die beiden Bogen Wolken gewesen sind. Ebenso wenig kann es sich um ein Nordlicht handeln, da der Mittelpunkt der Bogen von Anfang an nicht im Norden, sondern im Nordwesten lag und da ferner die Erscheinung durchaus ruhig und gleichförmig verlief, ohne jede Zuckung.

Die Vermutung liegt nahe, daß hier der Schweif des Halleyschen Kometen gesehen worden ist, der um diese Zeit die Erde nahezu gestreift haben kann."

Astron. Nachr. Nr. 4414, S. 365—369, 1910.

. . . In the printed observations that I have seen so far, of the appearance of the tail on the mornings of May 18 and 19 there seems to have been one part of the phenomenon visible here that has not been mentioned elsewhere. I refer to a luminous appearance, below the beam of light, that extended to the southeast horizon. The conspicuous feature of the display on the 18th was a rather broad shaft of light of nearly equal brightness throughout its width, with rather definite edges. This beam of light very much resembled a search-light. It stretched obliquely from the eastern horizon to the Milky Way in Aquila (a length from the comet's head of 107°), where it was lost. Between 2^h and 3^h a.m. it was a striking object. Near α Andromedae its width was about 4° or 5° and it tapered somewhat towards the end. Between ζ and γ Pegasi is was brighter than any portion of the Milky Way. Beyond ε Pegasi it was fainter — perhaps only one-fourth as bright. The star 71 Aquilae (Proctor's Chart) was exactly in the axis. ϑ Aquilae was exactly on the north edge. ζ Pegasi was on the north edge and slightly immersed in the beam of light, while γ Pegasi was about one-half a degree north of the axis. From this description, it will be seen that the object did not lie in the ecliptic, but made a considerable angle with it. This beam of light in all the accounts of Halley's comet on the mornings of the 18th and 19th of May seems to have been referred to as the "tail"; and I have seen no mention of any other phenomenon cennected with it.

Above this ray — to the north — the sky was free from any streamers or illumination of any kind. Below, to the southeast horizon, however, appeared to be a fainter diffused light with a comparatively dark space between it and the great beam, as if the beam were a brighter part only of the tail, while the main portion, which involved the ecliptic, extended to and below the horizon, and was separated from the brighter and narrower part by a rift — such as is frequently seen in comet's tails. The next morning showed this to be a real phenomenon belonging to the comet.

On the morning of May 19th, the ray of light had sensibly shifted to the north and was more tapering, and only about one half as bright as on the 18th. It was estimated to be as bright as the Milky Way. It southern edge was quite definitely defined, while the northern side was much diffused, especially near γ Pegasi, where it was some ten degrees wide. It could be traced to the Milky Way in Aquila (a length of 120°). I could not be sure of its extension beyond the Milky Way. It widened out very greatly near the eastern horizon, where it seemed to join a general illumination stretching along the horizon for some distance. The general brightness of the beam was estimated to be one-half that of the 18th.

At 2^h20^m a.m. ϑ Aquilae and ζ Pegasi were in the axis, while γ Pegasi was inside its southern edge (about 3° within the edge. The general illumination below to the southeast horizon was more definite at its upper edge). Between it and the south edge of the beam, the sky was dark. The upper edge of this broad illumination was roughly bounded by the stars β and γ Piscium and η and ζ Aquarii, and it joined the beam near γ Pegasi. This broader portion of the tail was much less bright than the great ray. The phenomena described here were seen by the other observers then present . . .

E. E. Barnard, Astron. Nachr., Nr. 4431, S. 229/230, 1910.

May 20; observations see Fig. 3, No. 40 and 41

May 20. — (a) Plate XI, No. 40. — Comet first seen at 5 h.40m., or 13 mins. after sunset. S. branch of tail is much the brighter. Bushy tail, nucleus 6″ diameter. Great sheath envelopes head.

(b) Plate XI, No. 41. — The comet was picked up at about sunset, but no observations could be made until the sky had darkened considerably. The nucleus was yellow 3, shading off to yellow 4 in the streamers.

May 20. — Plate VII, No. 31. — Exposure from 4h.21m. tot h.39 m. First photograph after transit. Moon in sky with faint twillight. The diameter of the head through the nucleus in nearly 30′. Three broad envelopes visible. Very little tail seen owing to fogging of plate.

Transvaal Observ., Circular No. 4, 11th July 1910, p. 26/29.

The nucleus looks small and sharp, about 6″ in diameter (C 2248, May 20.678). There are sharp curved jets in the direction toward the Sun, but no comparison stars for measurement. . . .

WS 29, May 20.691, shows a sharp curved jet in the tail whose motion it is possible of follow for several days. Unfortunately there are no comparison stars in the neighborhood and direct measurement of the position of the tip is impossible. Its motion is in all respects analogous to that of May 12—16, so that $(1' - \mu)$ must be nearly 1.

N. T. Bobrovnikoff, Publ of the Lick Observ., Vol. X VII, pt. 2, p. 374, 1931 T

Telegr. aus Cambridge, Mass., 1910 Mai 20 $12^1/_2{}^h$ früh.

Campbell telegraphs Comet Halley visible eastern sky this morning. Tail at least 140°, lags far behind radius vector, Earth will probably not pass through tail.

Pickering, Astr. Nachr., Nr. 4414, S. 369, 1910.

May 21; observations see Fig. 3., No. 42 and 43

Telegram aus Cambridge, Mass., 1910 Mai 21 $7^3/_4{}^h$ vormittags.

Campbell telegraphs: comet's tail in east Friday morning extending through Aquila 1° further north and one thirds as bright as yesterday.

Pickering.

Telegram aus Athen, 1910 Mai 21 nachm.

Queue de la comète Halley observée hier matin exactement même position derniers jours, hier soir aussi dans linette paraissait dirigée vers soleil, par suite grande inclinaison ou courbure. Passage de tête n'a manifesté aucun phénomène sensible.

Eginitis.

Telegram vom Observatorium Johannesburg, 1910 Mai 21 vormittags.

Halley evening tail 19°, also branched tail visible east this morning.

Astron. Nachr. Nr. 4414, S. 369/370, 1910.

The Nucleus is large and fuzzy, about 10″ in diameter. The star trails are sharp, so that the fuzzines of the nucleus is not due to defective guiding. This is interesting because the Johannesburg observers say that the nucleus was sharp, very small and of a yellow color. On Helwan plate HR 21, May 21.240, taken almost simultaneously, the nucleus hat the same appearance as on the Crossley plates, with diameter 10″. Probably the dense envelope of 10″ diameter consiste chiefly of CN hardly observable with the eye but strong photographically.

There are several sharp emissions. The dark gap in the *np* quadrant mentioned by the Johannesburg observers is well seen on the Helwan plate, but it is unquestionable brighter than the background of the sky around the comet.

Objective prism spectrogram, OP 11, May 21.479, has a sharp nuclear line 1′ wide. The CN IV band has a bright condensation of 1′.3 diameter and a fuzzy envelope of 15′ symmetrical on both sides of the nucleus.

N. T. Bobrovnikoff, Publ. of the Lick Observ., Vol. X VII, pt. 2, p. 374, 1931.

May 21. — Plate VII, No. 32. — Exposure from 4h.36m. to 4h.59m. Condensed stellar nucleus with well-marked but not symmetrical envelopes, i.e. the intensity of the envelopes more pronounced on the North edge than on the South.

Transvaal Observ., Circular No. 4, 11th July, 1910, p. 26

May 22; observations see Fig. 3, No. 44 and 45

The Crossley plates of May 22 show several sharp nuclear jets. They have slight condensations on their tips as if formed by instantaneous copious eruption. They look like secondary nuclei but are much fainter than the principal one. It is possible to distinguish them also on the Helwan plate of May 22.247 (HR 22). The measurement of p.a. is uncertain as the stars have long trails. Table 50 gives the p.a. and length of the jets:

Table 50

		Length		
		Point (1) HR 22	Point (2) C 2254	Point (3) C 2257
Jet ppa.	(C 2256)	May 22.247	22.678	22.733
a	293°	60″	73″	82″
b	312	30	58	77
c	329	30	43	54
d	340	30	43	54
e	213	—	27	27

It is evident that there must be considerable acceleration in the motion of jets *a*, *b*, *c*, *d*. Jet *e* is apparently an altogether different formation; it looks short and stubbly and shows no appreciable motion.

There is a noticeable deformation in the four former jets during the short interval between C2254 and C2257. Jet *a* clearly shows a discontinuity as if its tip was bent and partly broken off.

The p.a. of the prolonged radius vector is 104° for this date, so that jet a is directed *toward* the Sun. The cometocentric velocity can therefore be roughly computed. Taking into account the fact that the phase-angles for these three dates were 125°, 121°, 121°, and $\log p = \bar{1}.2465, \bar{1}.2701, \bar{1}.2726$ we obtain the linear distance of the tip of the jet from the nucleus 626, 769, 869 $\times 10^{-7}$ A.U. This gives the average velocity between the points (1) and (2) 0.057 km/sec and between points (2) and (3) 0.311 km/sec. The logarithm of acceleration in A.U. per day is therefore $\bar{4}.694$. Comparing this with $\log k^2 = \bar{4}.471$, we find that tip of jet *a* was moving toward the Sun under a force 1.67 times the force of gravitation. The excess over the force of the solar attraction is 0.67 in gravitational units k^2.

It is hardly conceivable that the Sun attracted the matter in the jet with a force greater than the force of gravitation. We may therefore conclude that the jet was under the repulsive force of the nucleus of the comet. Just what the magnitude of this force was is difficult to say. The value obtained is the result of a joint action of (1) the repulsive force of the Sun directed toward the nucleus, (2) the repulsive force of the nucleus directed away from the nucleus, (3) the attractive force of the nucleus directed toward the nucleus and (4) the attractive force of the Sun directed toward the Sun. Under these circumstances the most we can safely is that the nucleus exerted appreciable repulsive force on the ejected matter in jet *a*. This is in accordance with the motion of matter in jets *b*, *c*, *d*, also directed toward the Sun. All of these showed acceleration. It will be recalled here that a similar phenomenon of the same approximate magnitude was found in the motion of a jet directed toward the Sun on May 8—10. We shall see further that the motion of matter toward the Sun makes it possible to estimate the mass of the comet.

The nucleus on this date was small and sharp, from 3″ to 5″ (C2254 and C2255). It clearly shows a circular envelope 10″ in diameter.

The Johannesburg drawings of this date show jet *e* very prominent in contrast to the Crossley photographs. The nucleus is described as yellowish.

N. T. Bobrovnikoff, Publ. of the Lick Observ., Vol. X VII, pt. 2, p. 374/375, 1931.

May 23

The seven Crossley plates of May 23 show a remarkable phenomenon already descirbed by Curtis.[47] Two globular appendages near the nucleus developed within 50^m. The velocity of expansion was found by Curtis to be 0.88 km/sec.

C2259, May 23.679, has a large nucleus about 15″ in diameter. There are many strong jets of which five are the most prominent. Table 51 gives the p.a. of the jets on C2260, May 23.680. Reference to various observers are as follows: Battermann[48] at Königsberg; Hayn[49] at Leipzig, Voûte[50] at Leiden; H.C. Wilson[51] at Northfield.

Table 51

Jet	Battermann May 23.330	Hayn 23.344	Vo te 23.383	Wilson 23.646	C 2250 23.680
a				181°	187°
b				219	212
c	225°	230°	220°	219	226
d				258	253
e				301	310

The are many more observations of these jets but Table 51 suffices to show that the measurement of p.a. hardly certain within several degrees. Short-period oscillations of jet *c* are improbable. More likely we have to deal here with uncertainty in the estimate of p.a. . . .

Graff, at Bergedorf[55], May $23^d\!.37$, observed a sharp narrow emission in p.a. 229° (that is, jets *c* and *d*). At 23.33, he noticed „einige (3—5) blitzartige Zuckungen im ganzen Kometenkopf, die die Gesamthelligkeit für etwa $0^s\!.5$ um rund $0^m\!.5$ dämpften." The nucleus was round, 3″—4″ in diameter. Drawings by Flammarion,

May 23.355, and Quénisset, May 23.397, made at Juvisy[56] show approximately the same picture as the Mount Hamilton plates.

On this night a diffuse formation was moving in the tail. Its positions could be secured only on account of the eclipse of the Moon. . . . $(1 - \mu) = 56 \pm 42$.

[47] Publ. Astr. Soc. Pac. **22**, 117, 1910.
[48] Astr. Nachr. **187**, 27, 1910.
[49] Astr. Nachr. **187**, 26, 1910.
[50] Astr. Nachr. **187**, 347, 1911.
[51] Pop. Astr. **18**, 477, 1910.
[55] Mitt. Hamburger Sternwarte **5**, 62, 1913.
[56] Bull. Soc. Astr. France **24**, 282, 1910.

N. T. Bobrovnikoff, Publl. of the Lick Observ., Vol. X VII, pt. 2, p. 374/375, 1931.

— Mai 23. Im Äquatorial: Kern rund, 3″—4″ im Durchmesser. Schmale springbrunnenähnliche Ausströmung in der Richtung nach der Sonne zu im Pos.-Winkel 229°.2 (7 Einstellungen 9h.8m. Z. Bgd.). Die Schweifachse liegt nicht genau diametral gegenüber, doch ist eine Messung ihres Pos.-Winkels nicht mehr möglich. Der Kopf hat mindestens 5′—6′ im Durchmesser und erscheint sehr verwaschen, jedenfalls nicht scharf begrenzt. 1^h vorher, beim mikrometrischen Anschluß, beobachtete ich einige (3 bis 5mal) blitzartige Zuckungen im ganzen Kometenkopf, die die Gesamthelligkeit für etwa 0s.5 um rund 0m.5 dämpften. Im Opernglase finde ich den Kometendurchmesser durch Schätzung gegen Nachbarsterne etwa gleich 15′. Mit freiem Auge 10h.1 m. Z. Bgd. gleicht der Komet α Geminorum (PD 1m.94) und ist merklich heller als Mars und α Hydrae (HP 2m.16).

K. Graff, Astr. Nachr., Nr. 4539, S. 46, 1912

Mai 23. Heller Ausläufer vom Kopfe im Pos.-W. 230°, umgeben von Nebel.

Ç. F. Pechüle, Astron. Nachr., Nr. 4416, S. 399, 1910

Der am 23. Mai noch ziemlich große, nicht sehr scharf begrenzte längliche Kern zog sich rasch zusammen und blieb längere Zeit hindurch fast sternartig und rund, so daß sich sein Durchmesser mikrometrisch genau bestimmen ließ. Die Koma zeigte sehr weniges und verwaschenes Detail, das eine mikrometrische Einstellung nicht vertrug, und im Refraktor oder Sucher ließen sich die beiden um den Kern herumgebeugten Schweifansätze nur eine kurze Strecke verfolgen; der schwache Schweif verliert sich für das bewaffnete Auge schon bald auf dem nicht ganz dunklen Himmelsgrund. Die auf 0″.01 angegebenen Kerndurchmesser beruhen auf mikrometrischen Messungen bei Vergr. 207 (je 4 Einstellungen), die auf 0″.1 angegebenen auf Schätzungen; die Kernhelligkeit ist bei derselben Vergrößerung am großen Refraktor, Durchmesser und Helligkeit der Koma am Sucher beobachtet.

1910	M.Z. Straßb.	Kern			Koma		Bemerkungen
		Durchm.	Helligk.	Farbe	Durchm.	Helligk.	
Mai 23	$8^h.6$	6″.13		4°.5			Kern länglich, gemessen die lange Achse, PW = 53°.7 (4 Einstell.), (kleine Achse
26	9.7	2.15			15′		
27	9.1	1.95	$7^m.5$	2.5	15		Kernhelligkeit im Sucher geschätzt
28	9.0		9.2	4.5	8		
31	9.4	1.88	8.7	4	10	$5^m.5$	
Juni 2	8.8	0.5					
3	9.2	1.09		5		5.9	Kern sehr wenig länglich im PW 12°
8	9.5	2.3	9.6				Kern unregelmäßig begrenzt, zentrale hellere Kernscheibe von 20″ Durchmesser

C. Wirtz, Astron. Nachr., Nr. 4446, S. 93/94, 1911

May 23. — (a) Plate XI, No. 46. — Nucleus eccentric.
(b) Plate XI, No. 47. — Nucleus small, elongated 150°—330°, yellow = 2. Greatest glow on S.p. side of head. N. tail brighter.

Transvaal. Observ., Circular No. 4, 11th July 1910, p. 26.

Telegramm aus Treptow bei Berlin, 1910 Mai 22, $11^3/_4{}^h$ abends.
Sah heute $9^1/_2{}^h$ etwa 1° südlich vom Komet Halley ein kometenartiges Begleitobjekt mit halb so hellem Kern.

Archenhold.

Auf telephonische Anfrage teilt Prof. Schorr, Bergedorf, mit, daß auf der dortigen Sternwarte von vier Beobachtern, die am Abend des 22. Mai bei vorzüglicher

Luft den Kometen beobachtet und auch die Umgebung durchmustert haben, ein Begleitobjekt nicht gesehen wurde. Auch am Abend des 23. Mai waren die Nachforschungen vergeblich.

Kb.

Auf telegraphische Anfrage teilt Prof. Wolf, Mai 23, $11^{1}/_{2}^{h}$ abends, gleichfalls telegraphisch mit:

Auf 10 gestrigen Aufnahmen von Helffrich und Lorenz, welche Sternspuren 10. Größe zeigen, ist kein sekundärer Komet vorhanden.

Wolf.

Astron. Nachr., Nr. 4414, S. 370, 1910

May 24; observations see Fig. 3, No. 48

Bedeckung des Sternes $8^{m}.5$ AG Lpz I 4615 durch den Kern des Halleyschen Kometen

Bei der Vorbereitung einer photographischen Aufnahme des Halleyschen Kometen 1910 Mai 24 wurde um $8^{h}40^{m}$ M.E.Z. bei ziemlich heller Dämmerung im Gesichtsfeld des 12-zöll. Heydeschen Refraktors bemerkt, daß der Kometenkopf sich unmittelbar auf den Stern $8^{m}.5$ AG Lpz I 4615 zu bewegte. Kurz vor 9^{h} war es ersichtlich, daß eine zentrale Bedeckung des Sternes durch den Kern des Kometenkopfes erfolgen würde. Da der Stern von der der Sonne abgewendeten scharf begrenzten Kometenkernseite erreicht werden mußte, war zu erwarten, daß der Moment des Eintritts des Sternes in den Kometenkern außerordentlich scharf beobachtet werden könnte. In der Tat verschwand der Stern plötzlich, nachdem er kurz vorher unmittelbar die Grenze des scharfen Kometenkerns berührt hatte, wobei einen Augenblick das Sternscheibchen an der Grenze des Kometenkerns zu haften schien. Nachdem während $28^{s}.1$ der Stern innerhalb des sehr hell leuchtenden Kernes selbst vollkommen unsichtbar geworden war, erschien er in den der Sonne zugewandten Ausstrahlungen der leuchtenden Kometenmaterie wieder als ein absolut scharfes bläulich gefärbtes Pünktchen, dessen Helligkeit, soweit beurteilt werden konnte, nicht von der Helligkeit des Sternes vor dem Eintritt in die Kometenmaterie abwich. In dem Maße, wie der Kometenkern sich vom Stern entfernte und der letztere in immer weniger lichtstarke Regionen der nach der Sonne zu gekehrten Kometenausstrahlung gelangte, trat der Stern wieder deutlich aus der weniger leuchtenden Umgebung hervor, bis er schließlich aus dem fast dunklen Hintergrund des Himmels aus dem Kometen heraustrat. Während der Zeit des Durchganges durch die Ausstrahlung schien der Stern weder zu flimmern, noch war seine äußerst scharfe punktförmige Form irgendwie verändert, auch die Farbe schien keine Änderung zu erleiden. Auch sonst wurden keine irgendwie bemerkbaren Phänomene beobachtet.

Die Zeiten waren folgende:
Eintritt des Sternes in den Kometenkern $9^h0^m12^s.4$ M.E.Z. Wiedersichtbarwerden des Sternes in der leuchtenden Kometenausströmung $9^h0^m40^s.5$.

Der Eintritt des Sternes konnte mit äußerster Schärfe beobachtet werden. Die Uhrzeiten sind etwa auf $0^s.5$ unsicher, weil der Gang unserer einfachen Pendeluhr nicht vollkommen regelmäßig ist. Vor der Beobachtung und am nächsten Morgen nach der Beobachtung wurde der Stand der Pendeluhr mit der Zeit der Königlichen Sternwarte telephonisch verglichen. Vergrößerung 65fach.

Eine Aufnahme des Sternes in unmittelbarer Nähe des Kometenkopfes, die gleich darauf vorgenommen wurde, gelang nicht, da der Stern sich schon bei einer Exposition von etwa 6 Minuten nicht mehr von der stärker chemisch wirkenden leuchtenden Kometenausstrahlung abhob. Später trat Nebel ein.

Photogr. Sternwarte der Kgl. Technischen Hochschule, Berlin, 1910 Mai 25.

A. Miethe.

Astron. Nachr., Nr. 4417, S. 15, 1910.

Die Bedeckung des Sterns AG Lpz II 4615, über die Herr Miethe in A.N. 4417 berichtet, konnte von keinem der hiesigen Beobachter mit voller Konzentration verfolgt werden, doch ergeben die Wahrnehmungen, daß der Komet um Bruchteile einer Bogensekunde nördlich von dem Stern vorbeiging und daß ein eine größere Reihe von Sekunden dauerndes Verschwinden des Sterns nicht erfolgt ist. Aus Herrn Miethes Beobachtung, die sich aus der verhältnismäßig geringen von ihm verwandten Vergrößerung erklärt, darf also nicht auf eine starke Absorptionswirkung des Kometen geschlossen werden.

K. Schwarzschild.

Astron. Nachr., Nr. 4425, S. 140, 1910.

. . . straight jets *c* and *d* in p.a. 240° underwent a great change. They are now considerably curved and extend into the tail on the north side.

The Cordoba photograph of May 24.454 shows the curved jets, but there is nothing unusual in p.a. 60°. There is no doubt that the globular formations have vanished.

Barnard, on May 24.615, noticed the white color of the nucleus and its rapid increase in size (about five times) during the exposure. It will be recalled here that the color of the nucleus before the explosion of May 23 was decidedly yellow.

N. T. Bobrovnikoff, Publ. of the Lic Observ., Vol. X VII, pt. 2, p. 378, 1931.

May 25

The Crossley plates of May 25 show a sharp, short jet issuing from the nucleus in p.a. 109°, that is, exactly on the prolonged radius vector. Its lower end is almost wholly detached from the nucleus indicating that the emission was of short duration. . . .

The value of the repulsive force comes out $(1 - \mu) = 0.4$, but it is rather uncertain.

The heliocentric velocity corresponding to R_0 is 27.8 km/sec. The cometocentric velocity from points (2) and (1) is 1.60 km/sec., and from (3) and (2), 1.83 km/sec. The acceleration in reference to the nucleus is therefore wholly due to the elliptic orbit of the nucleus.

Gale[59], at Sydney, Austrialia, on May 24.9, observed a secondary nucleus "30" behind the principal nucleus connected with it by a bright of light." This was evidently the jet described above but the observer unfortunately did not give the exact time.

Other jets on the Crossley plates are faint and there are no distinct general envelopes. The nucleus is from 7″ to 10″ in diameter surrounded by an irregular bright nebulosity 70″ in diameter.

[59] Jour. Brit. Astr. Assoc. 21, 121, 1910.

N. T. Bobrovnikoff, Publ. of the Lick Observ., Vol. X VII, pt. 2, p. 378/379, 1931.

May 25. — Plate VII, No. 35. — Exposure from 5h.17m. to 5h.34.5m. Stopped by cloud. Stellar nucleus with symmetrical envelopes. Tail composed of numerous rays, the side ones soon fading away and leaving only the brighter central rays.

Transvaal Observ., Circular No. 4, 11th July 1910, p. 26.

Mai 25 wurde er mit freiem Auge von der Terrasse sichtbar 8^h10^m M.E.Z. und bis 10^h beobachtet. Gesamteindruck von 9^h bis 9^h15^m am schönsten. Kern gelb mit Nebelhülle; Schweif mit freiem Auge verfolgbar bis unter Regulus. Sofort nach Mondaufgang Abnahme der Sichtbarkeit des Schweifes und 10^h5^m undeutliche gelbe Nebelmasse im Dunste des westlichen Horizontes. Am Pentaprisma der kleine recht glänzende Kern scharf hervortretend gegen die schwache Nebelhülle. Am 4-Zöller mit 25-facher Vergrößerung Kern eine runde helle scharf begrenzte kleine Scheibe, intensiv orangefarbig; Nebelhülle fächerförmig, etwas über 180° auf der von der Sonne abgewandten Seite; auf der Sonnenseite keine Nebelhülle sichtbar. Im 6-Zöller bei 73-facher Vergrößerung dieselbe Erscheinung mit noch etwas schärfer ausgesprochenem Kern.

O. Kučera, Astron. Nachr., Nr. 4417, S. 13, 1910.

May 26

The Crossley plates of May 26 show another outburst from the nucleus. C2274, May 26.691, has a strong jet in p.a. 111°, 97″ from the nucleus. Its tip *a* is strongly condensed; a similar condensation *b*, 40″ diameter, is in p.a. 94°; its center is about 60″ from the nucleus. Both condensations *a* and *b* send out strong streamers *A* and *B*. Condensations *a* and *b* are receding from the nucleus and with them recedes the origin of the streamers *A* and *B*. This is a very convincing demonstration of how nuclear jets generate the fine streamers of the tail. It was possible to measure the position of *a* on five dates (see Table 58). The motion proves to be of the same character as of the jet on the preceding date, with nearly the same heliocentric velocity. . . .

Motion was therefore unaccelerated in reference to the Sun ($1 - \mu = 1.0$). The residuals are as follows:

The heliocentric velocity $R_0' = 27.0$ km/sec. The cometocentric velocities are; point (2)—point (1), 0,48 km/sec; (3)—(2), 0.49 km/sec; (4)—(3), 0.90 km/sec; (5)—(4), 1,46 km/sec. Here again the apparent acceleration in reference to the nucleus is due to the elliptic motion of the nucleus.

Other features of interest on the plates of May 26 are: a fan of 150° opening, with the axis in p.a. 292°, that is, nearly coinciding with the direction toward the Sun (p.a. 290°). It is 80″ long on C2274 (May 26.691). The edge of the fan in p.a. 217° is slightly curved into the tail; the edge in p.a. 7° is straight. This fan is mentioned by Hayn[60], on May 26.386, in p.a. 282°.

The nucleus seems to be elongated in the direction of the radius vector. Directly opposite to the Sun is a new short jet. It is sharp and bright on C2274 (May 26.691) but is totally absent on C2272 (May 26.681) and C2273 (May 26.684). Its length on C 2274 is 10″, which was attained in a few minutes. On C 2276 (May 26.762) it is about 40″ long. This gives an average velocity between these two points 1.10 km/sec. This jet dissipated very rapidly. There are no traces of it on the photographs of May 27.

The existence of persistent jets in the direction of the prolonged radius vector is remarkable. The can best be described as a series of explosions in that direction.

This also shows that the nucleus could not have a rapid rotation of several days duration unless its period was very nearly one day. However, a period of one day is not in agreement with observations of other cometary details, like the fan of this date described above. We have seen that it did not change its p.a. appreciably in $0^{d}.305$.

There are at least two spherical halos on C 2275 (May 26.726); the inner, brighter one has a diameter of 6'.5, the outer 12'.5. Both have the nucleus in their center. These halos are well defined on the Helwan plate of May 26.299; the inner halo is 5'.5 in diameter and the outer, 9'.0. There seems to be no doubt of expansion during the interval of $0^{d}.427$ with the average velocity, for the inner halo, 0.2 km/sec and for the outer, 0.7 km/sec. Of course the value of the velocity of expansion for the outer halo is less reliable than that for the inner.

The vertex of the envelope coincides with the inner halo, about 3'.3 from the nucleus.

Spectrogram W 222, May 26.726, shows the three principal knots CN IV, C III, C IV. The continuous spectrum is 36″ wide and has a still narrower central line probably corresponding to the nucleus. The core of CN IV_1 is about 1'.7 in length corresponding roughly to the emission fans. The core of CN IV_2 is only 1'.2 in diameter. There is a halo of 5′ diameter around CN IV and C IV corresponding to the inner halo of the direct photographs. The images of CN III and C V are 1′ in length and are situated on one side of the nucleus corresponding to the ejection fan toward the Sun. The spectrogram is too weak to show the CO^+ images and the outer halo.

[60] Astr. Nachr. **187**, 25, 1910.

N. T. Bobrovnikoff, Publ. of the Lick Observ., Vol. XVII, pt. 2, p. 379, 380, 1931.

One striking fact that was noticeable when the comet was bright in the evening sky, especially noticable on and about May 26, was that it had a nucleus within a nucleus. To the naked eye, there was a bright stellar nucleus of the 2.7 magnitude (= to that of δ Leonis). With field glasses the nucleus was small, but of sensible diameter, and of a beautiful bluish-white color; it was surrounded by a much fainter, hazy nebulosity, which ran out to form the tail — the view being rather an intensification of that with the naked eye. This naked-eye and field-glass "nucleus" was not the true nucleus. In the 5-inch guiding telescope a small planetary nucleus of the 8 or 9 magnitude was seen in a denser nebulosity. It was very well defined and very yellow. About this date, therefore, naked eye and telescopic observations of the nucleus would refer to two entirely different things of exactly opposite colors. That which formed the nucleus to the naked eye was simply the small denser nebulosity about the real nucleus.

E. E. Barnard, Astron. Nachr.; Nr. 4431, S. 234, 1910.

May 27; observations see Fig. 3, No. 49

The Crossly plates of May 27 do not show many details. C2281, May 27.782, has a strong emission in p.a. 267°, about 1′ long and perfectly straight. The nucleus seems to be slightly elongated in this direction. A fainter curved (clockwise) emission is about 30″ long, in p.a. 355°. Another still fainter emission is in p.a. 202°. C2279, May 27.695, shows two halos. The inner is about 4.′5 in diameter and its center is desplaced 1.′5 from the nucleus toward the Sun. The outer halo is concentric with the nucleus and is about 15′ in diameter.

Objective prism spectrogram W 223, May 27.745, shows the CN IV band surrounded by a halo 15′ in diameter which corresponds to the outer halo on the direct photographs. There is no halo of this size around C IV although its inner core (diameter 4.′2, that is, corresponding to the inner halo) is much brighter than that of CN IV.

An interesting observation was made by D. Goodbeard of Philadelphia[61] on May 27.573 (approximate). The telescope is not stated. He says: "Nucleus well defined and as bright as a star of the 7th magnitude. A flash from the nucleus clearly seen. It was confined to a point of light which darted seemingly earthward. After an interval of several minutes, a moderately bright flare-up occurred and while it lasted the dense part of the coma around the nucleus presented a mottled aspect. The flare-up gradually subsided in two and a half seconds of time. Atmospheric conditions for seeing at their best." As an isolated observation this seems to be hardly credible but it reads very much like Graff's observation on May 23.

The Johannesburg observers give fot the diameter of the nucleus 1.″5. (May 27.247).

Beginning with May 25 there was a remarkable development of activity in the tail (namely, the sharper and narrower tail which I have called previously Tail I). It was full of condensations and distortions moving away from the nucleus. We shall consider here several cases in which measurement was possible. Figure 37, Plate 17, May 26.726 (WF 31), shows condensations A, B, C. They formed a spiral in the tail and were moving under approximately equal repulsive forces. Table 61 gives the result of measurement (times corrected for aberration).

. . . Condensation A . . .	$(1 - \mu) = 60.0 \pm 10.1$
. . . Condensation B . . .	$(1 - \mu) = 52.6 \pm 2.3$
. . . Condensation C . . .	$(1 - \mu) = 76.0 \pm 9.2$

Table 61

Point	GMT	1910.0						Remarks
		α	δ	α	δ	α	δ	
1	25.703	8h55m18s	+7°20′	8h57m29s	+7°5′	8h58m26s	+7°08′	WF 30, 32
2	.786	57 22	7 05	58 47	6 53			DH
3	26.274	9 09 32	6 4	9 10 56	6 00	9 12 08	5 71	HP 28
4	.448	13 36	5 46	16 26	5 32	20 34	5 16	Cor
3	26.274	9 09 32	6 4	9 10 56	6 00	9 12 08	5 71	HP 28
4	.448	13 36	5 46	16 26	5 32	20 34	5 16	Cor
5	.710	19 57	5 13	24 07	4 50	26 53	4 38	WS 33
6	.724	20 13	5 09	24 39	4 48	27 25	4 37	WF 31
7	.739	20 37	5 06	25 13	4 46	27 51	4 33	WS 34
8	27.260	34 09	4 04	40 14	3 26	46 50	2 54	J 36
9	.475	39 33	3 37	46 52	2 45			Cor
10	.504	40 03	3 33	47 43	2 43			S 14
11	.743	53 46	2 11					WF 32

We can find the approximate velocity of ejection from the formula $g = r' - R'$, that is, assuming that the ejection was directed toward the Sun. If we assume the moment of ejection of condensation *A* to be its passage through perihelion ($R' = 0$), then $g = 29$ km/sec. If ejection occurred at moment (1), then $g = 6$ km/sec. The closest approach of the hyperbola to the orbit of the nucleus would correspond to $g = 15$ km/sec. The velocity of *B* was evidently of the same order.

[61] Pop. Astr. **18**, 576,1910.

N. T. Bobrovnikoff, Publ. of the Lick Observ., Vol. XVII, pt. 2, p. 382—388, 1931.

May 28

Beginning with May 28, another group of condensations was moving in Tail I. It was possible to measure 22 points between May 28 and June 2. Figures 39 and 40, Plate 17, show the position of the group (marked D) on May 28.751 and May 29.748. It will be seen from these reproductions that this condensation was not well defined except for its lower end, which was measured. The results of measurement and reduction to the plane of the orbit are given in the following table (times corrected for aberration):

Table 73

Point	GMT	1910.0 Condensation				Nucleus		Remarks
		α	δ	R	w	r	v	
1	May 28.287	9h31m50s	+4°12′	1.0005	+80°06′	0.9938	+80°20′	HP 33
2	.472	34 52	4 00	.0050	80 07	.9967	80 32	Cor
3	.720	38 44	3 33	.0128	80 37			WS 37
4	.749	3927	3 31	.0137	8033			WF 33
5	.766	39 49	3 27	.0147	80 39	1.0013	80 52	WS 38
6	.812	40 20	3 22	.0160	80 46			H
7	.859	41 09	3 17	.0177	80 51			H
8	29.194	46 52	2 45	.0290	81 13	1.0076	81 18	J 37
9	.278	48 22	2 40	.0312	81 08			HP 36
10	.473	53 00	2 15	:0404	81 16			SZ
11	.495	53 53	2 12	.0417	81 11	1.0102	81 29	Cor.
12	.720	57 12	1 47	.0507	81 42			WS 39
	.746	58 13	1 45	.0518	81 35			WF 34
14	.764	58 51	1 42	.0530	81 34	1.0165	81 54	WS 40
15	30.244	10 05 21	0 58	.0714	82 19	.0237	82 22	J 38
16	.495	07 50	0 49	.0764	82 20			Cor.
17	.540	08 36	0 47	.0775	82 15	1.0283	82 40	S 15
18	.744	11 19	+0 34	.0839	82 20			WF 35
19	31.234	19 58	−0 09	.1046	82 26			J 39
20	.452	23 08	0 3	.1164	82 51	1.0375	83 16	SZ
21	.747	29 52	1 01	.1309	82 36	1.0468	83 51	WF 36
22	32.745	49 46	−2 46	.1894	82 47	1.0622	84 49	Z 16

Plotting *R* and *w* against time (fig. 75) we notice that there is a remarkable break of continuity between points (14) and (15). In other words, it is impossible to reconcile the observations up to point (15) with the rest, without large and systematic departure. On the other there is no doubt both from direct observation and spectral analysis that we have to deal with the same condensation from point (1) to point (22). It is evident that the condensation underwent some transformations between May 29.8 and May 30.2 which changed its orbit completely. I have computed its orbit in two sections: D, comprising points (1) to (14) and E, from points (15) to (22).

We see that on May 26.45 the coordinates of condensation A were almost identical with those of D. This gives us a right to suspect at least that the condensation of May 28—June 2 was a by-product of a much earlier one of May 25—27. It may have been ejected from the lower end of the spiral structure (that is A) of May 26 (see Pl. 17, fig. 37). In such case the same condensations has changed its orbit twice: from A to D and from D to E. The conditions under which A could be transformed into D are $G = 17°$, $g = 44$ km/sec, which are almost identical with those under which D changed into E.

If this deduction is correct then the condensation in the tail on May 26 split into two parts: one went on without much change in its orbit, giving A and B. The other changed its orbit completely and was approaching the nucleus for several days before it condensed into the definite shape of point (1) on May 28.

We have to look therefore for traces of D on the plates of May 26 and 27. Plate WF 31, May 26.726 (Pl. 17, fig. 37), shows a condensation halfway between A and the nucleus. Later, on May 27 (WS 35, WF 32), a condensation which resembles D is nearer the nucleus than on May 26. It is, of course, difficult to settle this question on account of the complex structure of the tail. The most that can safely be said is that observation does not contradict the idea of the formation of D out of A.

Another contributing factor in the formation of D may have been the jets of May 25.8 and May 27.0, studied previously. We have near point (1) of D the heliocentric distance of the first jet $R = 0.9975$, and of the second $R = 0.9948$, while for D, $R = 1.0005$, and for the nucleus, $r = 0.9938$. The true anomalies were approximately the same. This means that the jets were halfway between the nucleus and the condensation and might have contributed something toward its formation. Plate 17, figures 39 (May 28.751) and 40 (May 29.748), represent successive stages in the evelopment of condensation D. Jets of May 25—27 are shown in figure 74 near point (1) of condensation D.

We have seen that the last point of A does not belong to that orbit and that B must have experienced some transformation between the time it left the nucleus and the first point observed. The situation is as complex as that on May 11—13 when a condensation might have been formed in three different ways. This phenomenon may be an explanation of the frequently observed sudden formation of condensations in the tail when they cannot be traced to the nucleus. They may be formed by a concurrence of several streams of matter, resulting sometimes in vortices and spirals.

We shall now continue the description of the cometary phenomena near the nucleus on May 28. On this date and also on May 29 and 30 it was possible to measure the outlines of the envelopes and the axes of the tails. Their measurement will be discussed separately together with the general structure of the head and

C2283, May 28.697, is reproduced in figure 36, Plate 16. It has a very complicated system of streamers which issue directly from the nucleus and do not form envelopes. They intersect the envelopes without any disturbance. The streamers form a cone of 30°, approximately symmetrical with respect to the radius vector.

The plate shows a bright emission in p.a. 299°, and at least three envelopes: inner, middle, and outer, decreasing in brightness with recession from the nucleus. The outer and middle envelopes can be seen on the reproduction. There is also a faint circular halo 19′ in diameter.

Objective prism spectrogram W 255, May 28.753, gives the usual spectral images. Their dimensions are CN IV outer halo, 17′, corresponding to the circular halo on the direct photographs; CN IV core, 8′ (outer envelope of direct images 13′, middle 7′); C IV outer halo, 10′, and inner core, 5′, corresponding to the middle envelope. C + H images are 2′—3′ in diameter corresponding to the inner envelope (3′). Condensation D is well visible in the CO^+ images.

N. T. Bobrovnikoff, Publ. of the Lick Observ., Vol. X VII, pt. 2, p. 388—395, 1931.

Observations of Halley's Comet made on board s. s. "Majestic"

By *J. Jenkins*, Fourth Officier. (Communicated by Capt. *Campbell Hepworth.*)

During the last voyage of the above ship from New York to Plymouth a good opportunity occurred for observing Halley's Comet. It was on the evening of May 28, and the night was exceptionally clear and cloudless.

The comet's head was directly under α Leonis, and I should say about half as bright as that star in the centre of the nucleus. The tail at the end nearest the head would be about the same brightness as the bright parts of the Milky Way on a clear night. It followed a straight line from the head to α Virginis, and was visible by the naked eye for about half the angular distance from the head to that star, this making the visible part of the tail about 30° in length. The width of the tail at this part would be about 2° and at the head about 8′. These widths are very rough approximations, as it was only by noticing with the sectant the angular distance between two stars which appeared to correspond with the width of the tail that they were obtained. The luminosity of the tail gaded gradually and evenly as its distance from the head increased. The head, through the night telescope, appeared like a ball of white fluff, about twice the size of Jupiter as compared with that planet at the time, and at no time was there any sharp definition of outline of the head or tail. The brightness of the head and tail varied at times, something after the fashion of the northern Lights.

At 11^h55^m p.m. G.M.T. on May 28, the ship by stellar observations being in lat. 43°12′ N., long 42°5′ W., the comet's head bore S. 70° W. true; its angular distance as measured by sextant from α Leonis was 11°35′, from α Virginis 59°29′, from α Hydrae 12°43′. At 1^h32^m a.m. G.M.T. on May 29, ship's position being lat. 43°25′ N., long. 41°26′ W., the comet's head bore S. 89° W. true; its angular distance from α Virginis was 59°8′, from a Leonis 11°32′, from α Boötis 69°56′, from Jupiter 42°23′.

At 2^h10^m a.m. G.M.T. on May 29, ship's position being lat 43°30′ N., long. 41°13′ W., the comet's head set bearing N. 85° W. true. Height of eye 47 feet. The comet's head at this time was so faint, that the time of setting and bearing were obtained by angular distance from α Leonis and a previous comparison of bearing with that star, both the angular distance and the bearings altering so little in comparison that they served to help to get a very good approximation of the time of setting, and also the bearing. The comet appeared at its brightest at the time the first set of observations were obtained.

Meteorological Office, London, S. W.; 1910 June 6.

Monthly Notices of R.A.S. L XX, 8 (1910) 619.

May 29

The Crossly plates of May 29 show a very small nucleus, hardly 4″ in diameter. There is an ejection fan in p.a. 275° of 6:5 length. It is certainly a new development as it was not discernible on May 28. There is a fuzzy jet in p.a. 112°, that is, along the prolonged radius vector, about 2:5 long. A complicated structure of faint jets toward the Sun is difficult to measure. There are two halos which are quite distinct from the envelopes. The inner halo is 2:5 in diameter and is very bright. The outer one is 9:0 and is very faint.

Spectrogram W 226, May 29.754 (reproduced in Pl. 21, fig. 68), shows the tail CO^+ streamers very well. Condensation D is well visible in CO^+. The dimensions of the spectral images as compared with the direct photograph C2288 (May 29.733) are as follows: inner brighter core of CN IV, 7:4; outer nebulosity around the same band toward the Sun, 8:2 and into the tail, 27′; diameter of C IV, 8:2 of C V and CN III, 4:4; of C + H, 2:5. On the direct photograph the inner envelope along the radius vector is 7:8; the middle envelope extends into the tail for 26:0 and toward the Sun, 4:6, that is, it coincides with the CN IV bands. The ejection fan (the brightest part extending 40″ and the fainter 3′) corresponds to C + H, C V and CN III, as well as to the inner core of CN IV and C IV.

There is a bright jet near the CN IV and CN III bands extending 5′ in the direction toward the Sun (discernible on the reproduction). The C V band does not show any traces of it although the band itself is brighter than CN III.

This jet (or rather fan) is mentioned by the Johannesburg observers. According to Barnard the nucleus was not yellow but pale in color; diameter N to S was 2.′6.

N. T. Bobrovnikoff, Publ. of the Lick Observ., Vol. X VII, pt. 2, S. 395, 1931.

May 29. — Plate VIII; No. 37. — Exposure from 4h.34m. to 4h.50m. Large bright nucleus with central dark space behind. Irregular tail North half disappears, leaving only South part.

May 29. — The tail after the first 30° from the head extends as a very faint and diffused beam of light, somewhat curved towards Jupiter and N. of Spica. Stellar nucleus without jets, but there is a broad brush of nebulosity about 10′ long springing from the head and pointing sunwards. Tail formation near head much brighter on S. side (5h. G.T.)

Transvaal Observ., Circular No. 4, 11th July 1910, p. 26, 30.

May 30

Mai 30. Noyau 8″, angle de direction de la queue près de la tête de la comète 109°.

F. Gonnessiat, Astron. Nachr., Nr. 4453, S. 206, 1911.

C2290, May 30.689, shows two short curved jets (counter-clockwise) in p.a. 192° and 218°; otherwise the head is structureless. The nucleus is 4″ in diameter.

On spectrogram W 228, May 30.752, the outer halo around CN IV and C IV is 10′ in diamter corresponding to the outer halo on the direct photographs. The inner halo of both bands is 5′.5. The CN IV bands have the same distribution of intensity as the direct picture, namely the sunward side is brighter while the C_2 bands show just the opposite effect.

May 31; observations see Fig. 3, No. 50 and Fig. 4, No. 1

The comperatively simple structure of the head on May 30 became exceedingly complicated on May 31.

Innes, on May 31.253, describes the appearance of the nucleus: "Nucleus stellar and yellow, unusual jets and rays. A second slight condensation behind the nucleus." On his drawing a condensation with a sharp point connected with the nucleus by a thin thread is shown in p.a. 114°, that is, almost coinciding with the axis of the tail.

Eginitis,[62] at Athens on May 31.288, describes this jet as a very brilliant one brighter than the nucleus itself. It was composed of several straight lines 50″ long forming a fan of 60° opening.

The meaning of this description becomes clearer when we consider the Helwan photograph (H 40) obtained on May 31.279. A drawing made from this photograph is reproduced in figure 76a. There are four globular formations *a*, *b*, *c*, *d*. Of these *a* and *b* are slightly elongated and are so situated that their axes are in p.a. 117°.5. Their diameters are 15″ in this p.a. and 10″ in the perpendicular direction. The distance between their tips is 47″. The nucleus must be near the tip of *a*, but it is hardly distinguishable from the rest of *a*. Globular formations *c* and *d*

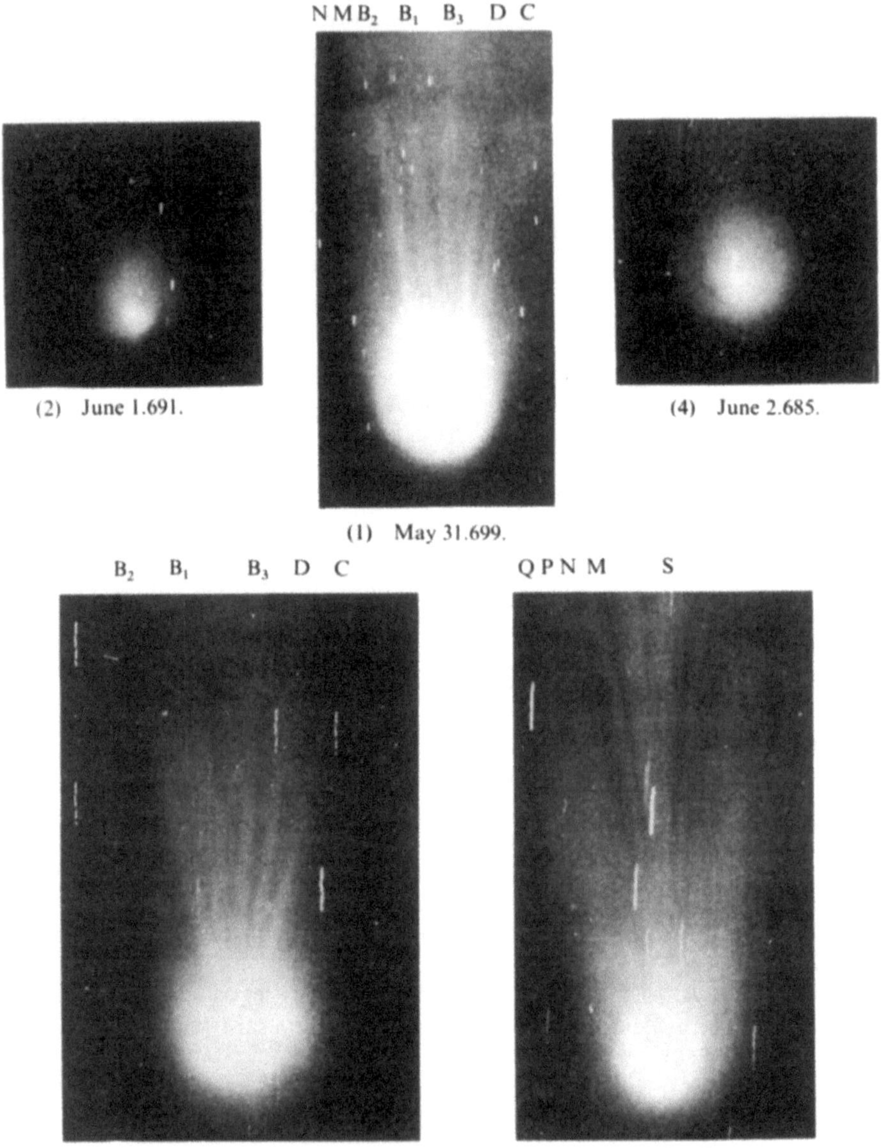

(2) June 1.691. (1) May 31.699. (4) June 2.685.

(3) June 1.729. (5) June 2.733.

Fig. 4. Halley's comet between May 31 and June 02, 1910; N.T. Bobrovnikoff, Publ. of the Lick Observatory, Vol. XXII, pt. 2, plate XVIII

have their centers in p.a. 6°0 and 58°0; their diameters are 17″. A circle of 40″ radius drawn from the nucleus (that is from the tip of *a*) is tangent to *c*, *d* and will intersect *b* near its center. The space within this circle is filled with luminous matter which is various degrees of brigthness. The globes are connected with each other by curved lines so that *a* and *b* form a dumbbell; *a* is joined to *d* as if matter were continually flowing in this direction.

A sharp curved line *k* separates all these formations from the rest of the head. Portion *e* between the eastern boundary of ad and the dumbbell formation is especially prominent. Fan *n*, 25″ in length with the axis in p.a. 326° is barely noticeable. In fact it was found only after comparing this plate with the Crossley plates of this date.

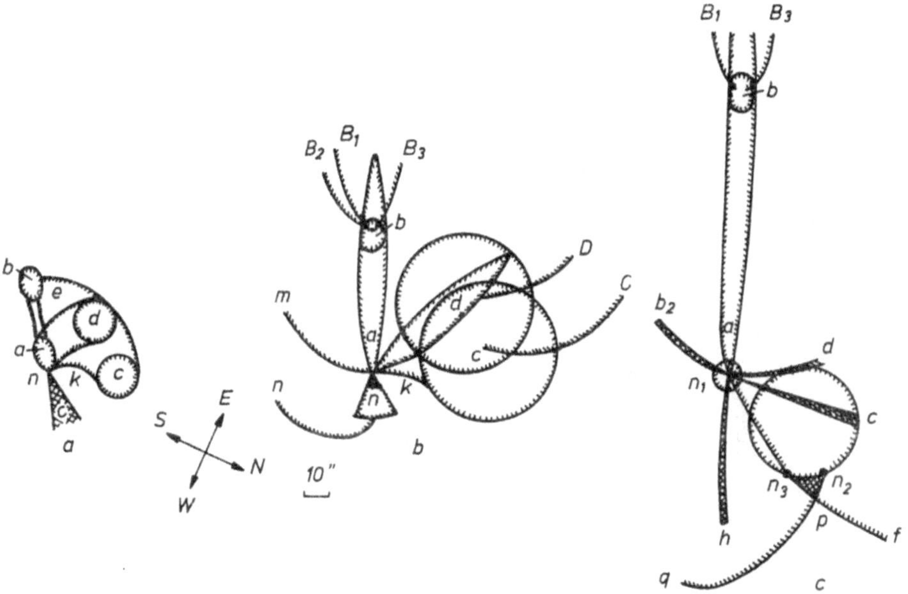

Fig. 5. The head of the comet on (a) May 31.279; (b) May 31.692, (c) June 1.678; after N.T. Bobrovnikoff; Publ. of the Lick Observatory, Vol. XVII, pt. 2, p. 396 (1931)

C2294, May 31.692, shows astonishing changes which occurred in the comet in less than half a day (see fig. 76b). The dumbbell formation *ab* has increased from 47″ to 90″ and its extremity farthest from the nucleus has lost its shape and diminished in brightness. Formation *a* has almost disappeared so that there is a gap between the ejected matter at *b* and the nucleus. Later the gap was again filled up which shows that ejection was intermittent. The centers of spheres *c* and *d* have receded from the nucleus; the spheres themselves have expanded and lost very much in brilliancy. They are recognized on this photograph with difficulty. The p.a. of the centers of the spheres have changed. Sphere *d* has a brighter central part of an oval shape. The axis of this oval coincides with the line joining the center of the sphere with the nucleus. The characteristic curve k has preserved its shape and is fully as sharp as on the Helwan plate. The nucleus is triangular with its axis in p.a. 302°. It measures 5″ along this axis, so that the real nucleus may have been even smaller than this. The sector of the nucleus has for its boundaries p.a. 276° and 328°. For about 19″ from the tip of the nucleus this sector is filled with luminous matter. The position angles distances, dimensions and brightness are as follows (for Helwan, May 31.279 and Mount Hamilon, May 31.692).

Table 84

Object Nucleus	Position Angle		Distance of Center		Diameter		Brightness	
	Helwan	Mount Hamilton	Helwan	Mount Hamilton	Helwan	Mount Hamilton	Helwan	Mount Hamilton
Nucelus	326°	302°.1					10	10
a	117.5	111.4					10	1
b	117.5	111.4	47″	90″			8	2
c	6.0	34.9	31	50	17″	57″	5	0.5
d	58.0	62.9	31	50	17	57	4	0.5
e							3	0

The brightness of the various formations was estimated on an arbitrary scale taking the nucleus as 10. Since the photographic plates were not of the same brand and since, furthermore, the Helwan photograph was studied on a slide reproduction, the figures give here cannot be claimed as accurate. A general decrease in brightness of the globular formations seems to be beyond doubt. This fact is confirmed by visual European observations.

From this measurements the velocity of recession of *b* from the nucleus comes out 0.43 km/sec; of *c* and *d* (projection on the prolonged radius vector), 0.19 km/sec; and the velocity of expansion of *c* and *d*, 0,20 km/sec.

The dumbbell formation *ab* is described by numerous European observers as a jet. Only a few observations will be mentioned here. We have seen that it was observed as early as May 31.253 at Johannesburg and May 31.288 at Athens and it is also present on the Helwan photograph of May 31.279. However, Borrelly[63] on May 31.364 describes the apparition of this jet during his observation. It was of the same brightness as the nucleus and was directed along the tail. Its length was 60″ and it persisted until the comet set. If this observation is correct, the jet suddenly flashed up and caught the attention of the observer.

The oscillation in the brightness of the jet with a gradual decrease seems to be well established. Eginites, on May 31.288, describes it as more brilliant than the nucleus. Hartmann,[64] at Göttingen on May 31.338, found the surface brightness of the nucleus 42 mkph (1 mkph = 10^{-6} standard candle) while that of the jet was only 9.0 mkph. Borrelly, on May 31.364, calls the jet as bright as the nucleus while Luizet,[65] on May 31.419, says that the jet was less brilliant than the nucleus.

This jet was situated approximately in the prolonged radius vector, so that the position of its tip could be easily measured with reference to the nucleus. Table 85 gives the results of measurement and the computed auxiliary quantities:

Table 85

Point	GMT	S	K	ξ	r	R	$w = v$	Remarke
1 May	31.279	47″	73°5	0.00011	1.03976	1.03987	83°24′8	HR 40
2	.288	50	73.5	12	.03990	.04002	25.3	Eginitis, visual
3	.364	1′00	73.3	15	.04105	.04120	29.7	Borrelly, visual
4	.695	1 34	72.4	24	.04617	.04641	49.1	C 2294; C 2295
5	.777	1 57	72.3	30	.04741	.04771	53.8	C 2297
6	1.678	4 56	70.3	80	.06130	.06210	84 45.3	MW 18
	.693	5 05	70.2	83	.06155	.06238	46.2	MW 19; C 2299
8	.723	5 31	70.2	90	.06203	.06293	48.0	MW 20; MW 21 C 2300
9	.760	5 54	70.1	97	.06259	.06356	50.0	C 2301
10	.779	6 07	70.1	0.00100	1.06288	1.06388	51.1	C 2302

There are altogether 14 observations. S is, as usual, the distance of the tip of jet b from the nucleus in angular measure, ζ is the corresponding linear distance in astronomical units, K is the phase-angle. All the necessary data were interpolated from Antoniazzi's ephemeris. The times were not corrected for aberration.

Besides the globular formations already described, three circular halos are discernible on the Helwan plate, HR 41, of May 31.279. They are denoted, beginning with the one nearest to the nucleus, as α, β, γ. These halos are well visible on C 2297, May 31.777. The remarkable formations near the nucleus and the similarity of the general aspect of the halos excludes all possibility of error in their identification. For the study of the expansion of the halos and the motion of their centers along the radius vector we have the following data:

Table 89

	Halo α		Halo β		Halo γ		Halo δ	
	Diameter	Distance	Diameter	Distance	Diameter	Distance	Diameter	Distance
May 31,279	1:9	+0:7	3:9	+0:0	6:8	0:0		
May 31.777	2.6	+0.3	6.0	+0.7	8.5	0.0	17:5	+2:6
Velocity of Expansion	0.16/km/sec		0.52 km/sec		0.42 km/sec			
Velocity of recession	−0.18		+0.36		0.00			

Halo δ is visible only on the Crosseley plate. In the column "Distance" is given the distance between the center of the Halo and that of the nucleus. The velocity of recession was counted as positive in the direction of the prolonged radius vector. We have then the surprising fact that while halo α was approaching the Sun β was receding from it. Hallo γ was apparently expanding symmetrically in all directions with respect to the nucleus.

On the Helwan plate of May 31.279 a pair of bright streamers begins at the nucelus. They form an angle of 15° and can be traced for about 20′ into the tail. They clearly issue from the nucleus, that is, they intersect at the nucleus or very close to it. The picture is entirely different on the Crossley plates of this date. The streamers of the Helwan plate have disappeared or else greatly diminished in brightness. Instead, there are a number of fine streamers issuing not from the nucleus but from various portions of the head.

On C 2295, May 31.699, the most prominent streamers *C* and *D* issue from formations *c* and *d* (see fig. 76b and Pl. 18, fig. 41). Strong streamers B_1 and B_2 are generated by the condensation near *b*. Streamer B_3 is formed by the same condensation but on the other side of the straight jet *ab*. Jets *m* and *n* produce two faint lateral streamers *M* and *N* visible only on the overexposed plates C 2296 and C 2297. Of all this streamers only *M* and *N* can be considered as nuclear.

The position angles of these streamers *at the source of their generation* are as follows (C 2297); B_1, 141°; B_2, 161°; B_3, 81°; *C*, 311°; *D*, 346°; *M*, 186°; *N*, 259°. The change in the relative intensity of the streamers is evident even within the short interval between the Crossley plates. The intensities were estimated on an arbitrary scale (see Table 90), assuming 10 for the brightnest streamer. They were regarded as a whole within 10′ from the nucleus.

These two plates are of comparable density. The enormous increase in brightness of streamers *B*, which can only be explained by the transformation of matter within the streamers, is striking. It may be noted here that the intensities of the streamers do not correspond to the intensities of their parent jets at all. Precise measurement is difficult but there is no question that *m* and *n* are much stronger than the rest of the formations, while *M* and *N* are the faintest among the streamers.

Table 90

Plate	Date	B_1	B_2	B_3	C	D	M	N
C 2295	May 31.699	2	9	3	5	10	11	
C 2297	31.777	10	8	8	8	9	1	1

The change in the intensity of the streamers in a short interval is strong evidence for real changes in the luminosity of matter in the tail, so often observed in caudal condensations. Motion of matter along streamers is obvious, but it is difficult to find well defined points for measurement. The only reliable measurement would be for a bend in streamer B_2. It moved about 40″ between C 2295 and C 2297 ($0\overset{d}{.}078$). This is sufficient to show that the change in the brightness of the streamers cannot be due to the motion of matter.

Comparing the WF plates on May 31.750, June 1.748, June 2.747, we notice that the lateral streamers *M* and *N* on May 31 disappear entirely at 30′ from the primary nucleus. The lateral streamers *C* and *D* can be traced up to 50′. The strong central streamer B_3 reaches to 1°02′ from the nucleus, but the tail is formed exclusively by the central streamers B_1 and B_2. On June 1, there is no evidence of *M* and *N*; streamers *C* and *D* reach to 1°30′ but are much fainter than the *B* streamers. Of these, B_3 extends 1°30′ but the tail is again formed by B_1 and B_2. On June 2, there remain only traces of *C*, *D*, B_3, but the central streamers B_1 and B_2 are very strong; they form a bifurcated tail.

From this follows that early in June Tail I was formed by the ejection from the nucleus along the radius vector (straight jet *ab*). This ejection formed a number of secondary centers of activity. These were not stable formations and soon dissipated. While they existed, they formed many fine streamers; this explains the peculiar aspect of the had on May 31. If on this date (see Pl. 18, fig. 41) the streamers be prolonged, they intersect far behind the nucleus. Next day when the secondary centers of activity disappeared the streamers intersect near the nucleus again (see Pl. 18, fig. 43), which shows that they were generated from the nucleus itself.

As has been mentioned before, the straight Tail I could be traced to bright streamers originating in condensation *b* on the prolonged radius vector. On Z 15, May 31.755, it is visible for a distance of about 24° from the head. It is full of condensations indicating great activity. At 9° from the nucleus it clearly separates from the structureless Tail II. It is impossible to trace the origin of Tail II but it must be due to irregular formations and ejections near the head. It is almost certin that the fine streamers have nothing to do with it.

Objective prism spectrogram W 230, May 31.755, shows three images of the CO^+ bands with the edges at λλ 4000, 4251 and 4543, the middle band being the strongest, as usual. All the tail images have a bright violet edge, probably due to streamer B_2. Other streamers can also be recognized in the tail images. Actual measurement of the spectrogram gave the length of B_2 as 38′; the beginning of a condensation in Tail I, 1°55′; a sharp bend to the south in Tail I, 2°37′ from the nucleus. These features correspond within the error of measurement to those of WS 43, May 31.724. There is no doubt that the straight Tail I consisted largely of CO^+. An interesting circumstance is that the band at C IV shows a tail apparently corresponding to Tail II. There is a double band of CO^+ near this wave-length (λ 4686 and λ 4714) but it is usually faint and cannot contribute much in that region. Moreover, there is not the slightest indication in this broad and diffuse tail image of the sharp details of Tail I (the band of CO^+ at λλ 4543—4569 affords a useful comparison). Another point is that the band of CO^+ at λ 4543 makes an angle of 2°.5 with the axis of the diffuse tail near C IV, exactly as on the direct photographs. This diffuse tail is fully as bright as brightest CO^+ band but shows no traces of streamers *B* and on the contrary is rather sharply defined on the red side. It is therefore likely that the two tails differed in composition. Tail II consisted either of carbon alon or of carbon with the admixture of small particles which give a sharp maximum near λ 4700 in their continuous spectrum.

Measurement of the diameters of the spectral images gave the following results:

Table 91

Object	Diameter	Direct photograph
CN IV outer spherical envelope	18′.9 (Center + 1′.6 from nucleus)	halo δ 17′.5 displaced + 2′.67)
CN IV inner sperical envelope	9.4	halo γ 8′.5
C IV spherical envelope	9.4	halo γ 8.5
C + H, CN III, C V outer envelope	4.4	halo β 6′.0
C + H, CN III, C V inner envelope	2.2	halo α 2′.6

These measurements indicate that halos α and β consisted of C + H, CN III, C V; halo γ, of CN IV and C IV; and halo δ, exclusively of CN IV.

It was impossible to recognize on this spectrogram the finer details of the head. There is α strong continuous spectrum about 2′ wide due either to the continuous spectrum of halo α or more probably to defective guiding. This of course obliterates all the nuclear details.

[62] Compt. Rend. **151**, 41, 1910.
[63] Bull. Astr. **27**, 365, 1910.
[64] Astr. Nachr. **185**, 233, 1910.
[65] Astr. Nachr. **190**, 9, 1912.

N. T. Bobrovnikoff, Publ. of the Lick Observ., Vol. XVII, pt. 2, p. 397—402, 1931.

Mai 31. On distingue.dans la tête de la comête trois aigrettes lumineuses situées à 120° environ les unes des autres. Une d'elles, dirigée sinsiblement dans l'axe de la queue ($P = 109°$), présente une condensation un pen plus étalée et moins intense que le noqau. A $10^{h}13^{m}.5$ (T.m. Paris) la distance des deux condensations est de 24″. — Juin 1
La condensation secondaire a disparu.

M. M. Luizet, Astr. Nachr., Nr. 4537,
S. 11/12, 1912.

May 31—June 02

El 31 de mayo, mientras efectuba la exposición (83 minutos) de una fotografia del cometa de Halley con el ecuatorial astro-fotografico de 38 cm del Observatorio Fabra, advertí la precencia de una nube ó bocanada de gases muy brillante, cuyo centro distaba aproximadamente un minuto de arco del núcleo y situada en la parte opuesta al Sol. Esta importante observación la transmití telegráficamente a'la "Centralstelle". Por otra parte, el clisé reveló la existencia de una inmensa proyección de gases situada hacia la parte media de la cola, de aspecto filiforme y cuyo centro de gravedad, con relativa aproximación, distaba un millón den kms. del núcleo.

El I de junio, el cielo estuvo brumoso; no obstante saqué una fotografia del cometa con 30 minutos de exposición, en la cual aparece un rayo de intensa luminosidad que surge del núcleo. Visualmente, con el ecuatorial, la citada nube

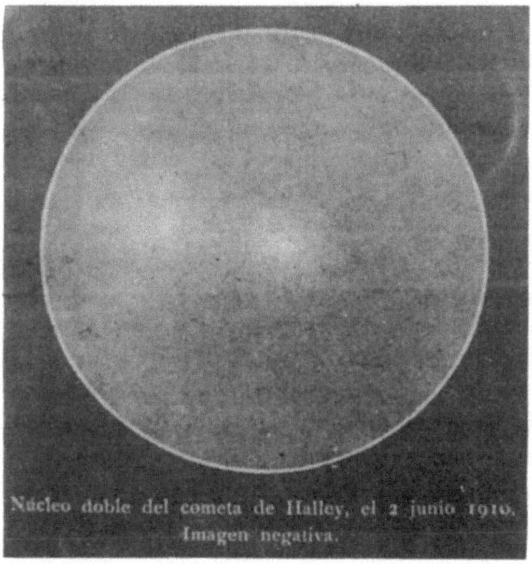

Fig. 6. Double nucleus of Halley's comet on June 02, 1910; J. Comas Sola, Astron. Nachr., Nr. 4421, 1910, p. 75

de gases era invisible; pero el 2 de junio, estando la atmósfera muy diàfana, observé de nuevo y con extrema facilidad la citada nube luminosa, aproximadamente á la misma distancia real del núcleo y en la misma posición con relación á la cola que el 31 de mayo. Dicha nube ó bocanada era muy brillante, y lo que es más notable, presentaba una condensación estelar en su centro. No se trataba sólo, por consiguiente, de una simple nube, sino de un desdoblamiento ó duplicación del nucleo. Acto seguido comuniqué la notica telegráficamente á la "Centralstelle". Del núcleo principal al secundario habia, el 2 de junio, una distancia angular de unos 40″, y el ángulo de posición correspondia, sensiblemente, con el de la cola, en el sentido opuesto, al Sol, según queda dicho. La diferencia de magnitud global entre los dos núcleos, rodeados de nebulosidad, aparecian enlazados por un ligamento nebuloso muy visible, en el cual, durante los mejores instantes de observación, pareciame distinguir una alineación de pequeñisimos puntos estelares. En fin, el clisé de este mismo dia, 2 de junio, obtenido con una exposición de 90 minutos, muestra un prolongado filete que surge de la cabeza y sigue la linea media de la cola. Ayer, 3 de junio, estando el cielo completamente cubierto, no me fué posible continuar estas observaciones.

Observatorio Fabra, Barcelona, 4 junio 1910

José Comas Solá

Astr. Nachr., Nr. 4421, S. 75/76, 1910.

May 31—June 04

Mit Bezug auf die Meitteilungen von Comas Solá auf Seite 75 des laufenden Bandes der A. N. über das Aussehen des Halleyschen Kometen am 31. Mai, I., 2. und 4. Juni teile ich kurz folgendes mit:

Am 31. Mai beobachtete ich (vergl. die Fig. auf der beiliegenden Tafel) auf dem Sonnenwendstein an einen Refraktor von 135 mm Öffnung und bei 25-facher Vergrößerung im Kopf des Kometen drei Nebenkerne, welche fast genau dieselbe Gruppierung zeigten wie die Zeichnung von Comas Solá am 4. Juni jedoch ohne die zwei schwachen Objekte, die hier mit *C* und (?) bezeichnet sind; es ist sogar die Reihenfolge der Helligkeit der Nebenkerne dieselbe. Es scheint mir wichtig dieses anzumerken, weil Comas Solá am gleichen Tage (31. Mai) nur einen Nebenkern feststellte. Meine Wahrnehmung wird bestätigt durch eine fast gleichzeitige Aufnahme von Dr. Krumpfholz am photographischen Refraktor der Wiener Sternwarte, die zwei der von mir gesehenen Nebenkerne unzweifelhaft erkennen und auch den dritten vermuten läßt. In Ermanglung geeigneter Hilfsmittel waren genaue Messungen auf dem Sonnenwendstein nicht möglich.

Am 1. Juni war auf dem Sonnenwendstein bei etwas dunstigem Himmel von den Nebenkernen nichts mehr zu sehen, hingegen war eine Lichtausströmung gegen die Sonne hin zu erkennen.

Fig. 7. Head of Halley's comet on May 31, 1910; the nucleus is splitted in three parts; I. Rheden; Astron. Nachr. Nr. 4431, 1910, p. 244

Am 2. Juni waren wegen ungünstiger Witterung keine Beobachtungen möglich.

Am 3. Juni war am Kometen nichts Auffallendes zu sehen; vielleicht war eine Lichtausströmung gegen die Sonne hin vorhanden. Eine gleichzeitige photographische Aufnahme in Wien zeigt einen guten fixsternartigen Kern und eine schwache Lichtausströmung gegen die Sonne. Von Nebenkernen war weder auf dem Sonnenwendstein noch in Wien etwas zu bemerken.

Am 4. Juni war auf dem Sonnenwendstein der Himmel sehr dunstig und der Komet nur schwach zu sehen; es ließ außer dem eigenen Kern und einer schwachen Nebelhülle nichts. Auffallendes wahrnehmen.

Der Widerspruch zwischen den Beobachtungen am 31. Mai in Barcelona einerseits und auf dem Sonnenwendstein und in Wien anderseits sowie zwischen den Beobachtungen vom 4. Juni in Barcelona und auf dem Sonnenwendstein ist ein so auffallender, daß zu seiner Klärung die Heranziehung der Wahrnehmungen anderer Beobachter wünschenswert erscheint. Von ganz besonderem Interesse wäre es aber den Zusammenhang zwischen den fast identischen Beobachtungen vom 31. Mai auf dem Sonnenwendstein und vom 4. Juni in Barcelona durch gute dazwischen liegende Beobachtungen herzustellen.

Wien, 1910 Juli 5.

J. Rheden

Astr. Nachr., Nr. 4431, S. 244, 1910.

Eine merkwürdige Erscheinung zeigt der Kopf des Kometen am 31. Mai und 2. Juni. Von dem scharfen Kern aus strahlte ein kurzer von der Sonne weg gerichteter Lichtarm, der sich von dem hellen Nebeluntergrund der Koma gut begrenzt abhob. Er hatte eine linsenartige Figur, doch ist es nicht sicher, ob die dem Kern zugekehrte Spitze jenen auch wirklich erreichte. Meinen Zeichnungen entnehme ich als Verhältnis Länge zu Breite den Wert 5:1. Am 2. Juni erschien das Gebilde ganz leicht gekrümmt, die konvexe Seite dem Horizont zugekehrt, und die breiteste hellste Stelle war etwas gegen das vom Kern abgewendete Ende hin verlagert; das Ganze gewann dadurch ein keulenähnliches Aussehen. An beiden Abenden befand sich unterhalb dieser hellen Stichflamme in der Koma eine auffallend dunkle Gegend, die am 2. Juni einem der Linsenachse parallel laufenden Kanal glich. Der Positionswinkel der eigenartigen Ausströmung wurde mikrometrisch gemessen und ihre Länge geschätzt wie folgt:

1910	M.Z. Straßb.	Länge	PW	$p_{\odot}$	$p - p_{\odot} - 180°$	Bemerkungen
Mai 31	$9^h.3$	$1'.0$	$119°.0$	$291°.8$	$+7°.2$	PW 4 Einstell.
Juni 2	9.4	0.9	116.3	292.2	+4.1	PW 8 Einstell.

Am nächsten Abend (Juni 3) sah man keine Spur mehr von dem beschriebenen Lichtstreifen, und die Koma hatte wieder ihr gewöhnliches gleichförmiges Aussehen.

Beobachtungen mit bloßem Auge gelangen an 6 Abenden. Sie erstrecken sich über Helligkeit, Schweiflänge und Schweifrichtung. Die Totalhelligkeit wurde durch Stufenschätzungen erhalten; die Größen der Vergleichsterne entstammen dem Potsdamer Generalkatalog, und der der Reduktion zugrundegelegte Stufenwert ergab sich zu $1^{st} = 0^m.083$. Die Schweifrichtung bestimmte ich durch Alignements, deren Nachmessung auf Sternkarten dann den formal auf Äqu. 1910 übertragenen Positionswinkel lieferte. Einigemale ist ein Feldstecher (Öffn. 36 mm, Vergr. 4-fach) zu Hilfe genommen, dessen Anwendung übrigens an der Art der Helligkeitsauffassung nichts änderte. Extinktion streng berücksichtigt.

Nimmt man die auf $r = 1$, $\Delta = 1$ reduzierte Totalhelligkeit konstant an, im Mittel $= 4^m.33$, so folgt der m. F. einer auf zwei Schätzungen beruhenden Helligkeit zu $\pm 0^m.10$.

Der Schweif war stets überaus schwach und in seiner vollen Länge nur bei indirektem Sehen wahrzunehmen; er schien gerade und schmal, am Ende 1°—2° breit oder auch, wie Mai 31, flach konisch zugrunde.

Bei der Betrachtung der Helligkeit und Schweiflängen muß man bedenken, daß der Komet von der Sternwarte aus im Westen über der Stadt stand.

C. Wirtz

Astron. Nachr., Nr. 4446, S. 95/96, 1911.

1910	M.Z. Straßb.	Beob. Helligk.	ZD ☿	Vergl.sterne	Red. auf Red. $r = 1, \Delta = 1$	Helligk.	Länge	Schweif PW	$p_{\odot}$	$p - p_{\odot} - 180°$
Mai	27 9^h40^m	$1^m.97$	73°	α, γ Leonis	$+2^m.42$	$4^m.39$	11°	108°.6	290°.6	−2°.1
Mai 27	9^h40^m	$1^m.97$	73°	α, γ Leonis	$+2^m.42$	$4^m.39$	11°	108°.5	290°.6	−2°.1
28	9 59	2.88	74	δ, ϑ Leonis	+2.13	5.01	4			
31	9 46	2.69	73	δ, ϑ Leonis	+1.50	4.19	22	105.2	291.8	−6.6
	10 3	2.50	76	γ, ζ Leonis	$+1^{+}50$	4.00				
Juni 3	10 9	3.24	78	15 Sextantis	+0.94	4.18	18			
	10 19	3.21	79	η, ϑ Leonis	+0.94	4.15				
5	9 56	3.91	76	15 Sext., ϱ Leonis	+0.60	4.51				
	10 0	3.69	77	η, γ Leonis	+0.60	4.29				
	10 0	3.69	77	η, γ Leonis	+0.60	4.29				
8	10 6	4.10	79	15 Sext., σ Leonis	+0.18	4:28				
						4.33				

Visual Observations of Halley's Comet and Comet 1910a.
Halley's Comet

Dr. A. A. Rambout

In consequence of the comet's proximity to the Sun, which rendered it a difficult object in these latitudes, observations were discontinued after the month of February 28, when it was just visible in the 18-inch in strong twilight. After perihelion it was first looked for, and seen, with the 10-inch Barclay in the early morning of April 25.

The positions of the comet, as deduced from visual observations, being obtained from circle-readings only (exept on May 22, as indicated by an asterisk), cannot lay claim to any great accuracy. They are, however, given in the following table, and may perhaps prove not wholly interest: —

Date	GMT	Appar. R.A. of Comet.	Appar. N.P.D. of Comet.	Instrument	Observers	Ref.
1910	*h m s*	*h m s*				
Apr. 24	16 28 52	23 50 18	82 72″	Barclay (10-inch)	R	*(a)*
May 20	8 24 12	5 25 53	70 58	Barclay (10-inch)	W. R.	*(b)*
20	8 51 27	5 27 12	70 58	18-inch	A.R. G.R.	*(c)*
22	9 15 30	7 18 52	75 44	18-inch	A.R. G.R.	*(d)*
22	9 32 15*	7 19 21′3	75 44 51	18-inch	A.R. G.R.	*(e)*
22	9 52 50	7 20 0	75 47	Barclay (10-inch)	H.B.	*(f)*
23	9 14 34	7 57 4	78 13	18-inch	A.R. W.	*(g)*

Observers' Remarks

(a) Apr. 24. Observed in strong twilight, using powers 45, 90, and 100. Nucleus only visible; magnitude 4, but image dull and of sensible diameter, apprxomately 8″. Too faint for the finder (2′7 inch) after 16^h30^m G.M.T.

Apr. 29. 14^h45^m — 15^h30^m G.M.T. In the 10-inch the comet appears about 3rd magnitude. Roundish, diffused, nebulosity of 40″ diameter. Nucleus stellar at times. An extension suspected in P.A. 290°—300°, but observations very difficult in rapidly increasing daylight. Thin haze over sky near eastern horizon. (A.R. and G.R.).

(b) May 20. Earliest opportunity for observing the comet since its passage to east of Sun.

The comet was detected in the 10-inch as a very faint nebulosity on a bright background of sunlit sky. In the finder the image was just discernible as a minute dull white patch.

(c) May 20. Rather brighter than ζ Tauri, 3rd mag.

(d) May 22. An extension or elong ation of the nucleus was noticed by Dr. Rambout in the direction P.A. 260°. This observation was corroborated by Dr. Cowell, who visited the observatory this evening.

(e) May 22. Position deduced from two transits of comet and star (Berl. A, A.G., 2772).

(f) May 22. From the nucleus of the comet a beam of light proceeds towards the Sun, but curves southwards.

(g) May 23. Nucleus elongated (3″ or 4″) in P.A. 225°. The coma extends in same direction like a fan of about 90° vertical angle to a length of 70″. (A.R.).
Nucleus seems forked at s.p. end. Coma extende for about 2′ in P.A. 45°. (W.).

May 23. 9^h15^m G.M.T. With the 10-inch Bàrclay telescope the outer edge of the coma is very indistinct, whilst the inner edge near the nucleus is sharply defined. Extending from the nucleus in the direction of the Sun is a narrow fan-shaped beam which curves round on either side, towards its brighter and the curving is more evident on the south side. (H.B.)
(Two photographs taken on this date — May 23 — confirm the above observations.)

May 24 9^h55^m G.M.T. Nucleus visible to the naked eye. Magnitude 3 · 8, and surrounded by nebulosity 5′ in diameter. Tail just discernible in twilight and moonlight, and could be traced to a distance of 8°. Inclination of tail to vertical, 55°, towards the south. (R.)

May 25 10^h G.M.T. The tail can be traced with the naked eye to a distance of about 30° (= distance between β Ursae Maj. and Polaris). Rather difficult, very diffuse, no definite boundary; practically straight throughout its entire length. (H.B.)

May 31 10^h15^m G.M.T. Tail just visible to naked eye in twilight. Angle with vertical, 60°—65°. Brigthness of head 3rd magnitude. (R.)

May 31 10^h30^m G.M.T. The tail is straight, and can be traced to a distance of 18°, or as far as φ Leonis, involving p^1 and p^2 Leonis. It is relatively bright for the first five degrees, bur rather narrow with diffuse margins. (H.B.)

June 2 — 10^h G.M.T. In the 18-inch, the horn extending from the nucleus at P.A. 120° is more marked than it was on May 31; the nucleus and comet generally are fainter than on that date. (A.R.)
Tail still visible to naked eye, clearly seen for five degrees and suspected five degrees farther. Magnitude of nucleus about 3. (R.)

Monthly Notices of R.A.S. LXXI, 2 (1911) 137.

June, 1910

June 01; observations see Fig. 5

MW 18, June 1.678, shows the nucleus fuzzy, 10″ in diameter. The head represents a highly complicated structure. The identification of the features on this date with those of May 31 is difficult, but there is nevertheless some degree of resemblance. The features which are supposed to be the same as on May 31 (fig. 76b) are denoted with the same letters on June 1 (fig. 76c). The streamers near the head are obviously analogous on both dates (compare Pl. 18, figs. 41 and 43). Unfortunately there is no Helwan plate of June 1 to assist in the identification.

The straight jet of May 31 has considerably decreased in brightness. It can be seen on Plate 18, figure 42 (parallel to the star trails). It is no longer straight but is slightly distored. It end *a* near the nucleus has dissipated entirely. The jet seems to be wholly detached from the nucleus. Its length is now 4′56″ in p.a. 109°. Formation *b* has receded from the nucleus to a distance of about 2′. It generates streamers B_1 and B_3 as on the preceding date. But the source of streamer B_2 is now at the nucleus, namely jet b_2. It is probably the old nuclear jet *m*. The p.a. of b_2 is 180° while the p.a. of *m* was 186°. Two very bright jets *c* (p.a. = 0°) and *d* (p.a. = 10°) are responsible for streamers *C* and *D*. If this is so, the sources of streamers *C* and *D* must have joined the primary nucleus again.

There is a remarkable globular formation of 48″ diameter with the center 38″ from the nucleus in p.a. 350°. It is probably distict from the globular formation of the preceding date. Two curved jets *f* and *g* are associated with the globe. Jet *g* begins on its periphery in p.a. 338° from the primary nucleus. Its p.a. at the origin n_2 is 282°. In p.a. 297° and 92″, counting from the primary nucleus, it reaches its vertex and curves back. Jet *f* starts in p.a. 322° on the same periphery. Its p.a. counted from this secondary center of activity n_2 is 340°. It reaches its vertex in p.a. 340° from the primary nucleus and 1′44″ from it. These two jets *f* and *g* are very fine, almost threadlike. Their sources n_2, n_3 and the point of intersection *p* form a luminous triangle. It will be noticed that jets *f* and *g* do not interfere with each other. They may be situated in different planes. It is possible that both are formed at point *p* and not at n_2 and n_3. Being very faint jets, *f* and *g* do not produce any streamers but disappear near their vertices. It is worthy of notice that a band joins the secondary center of activity n_3 with the nucleus n_1. There is also a faint straight jet *h*, opposite to *ab*.

As on May 31 the relative brightness of the streamers do not correspond exactly to the brightness of the jets.

The details near the nucleus on MW 18 are immersed in a halo roughly spherical about 4′ in diameter. Its center is displaced in the prolonged radius vector by about 40″. The outer fuzzy halo of 13′ diameter has the nucleus in its center.

The next plate, MW 19, June 1.691, gives some evidence of rapid changes in the head. Jets *g* and *f* are much more distinct; *f* seems to curve back into the globular formation making in a loop, but unfortunately a 14th magnitude star is involved in it. The globe expanded so much as to include point *p*. On MW 20, June 1.712 and MW 21, June 1.726, the space between *p* and the globe is clear again but the jets are diffuse and blurred.

It is possible to measure the motion of matter in the streamers between June 1.678 (MW 18) and June 1.726 (MW 21). The condensations are blurred and measurement was only approximate.

Table 92

Streamer	MW 18	MW 21	Displacement
B_3	647″	726″	79″
C	635	726	91
D	645	753	108
		Mean	93″

This indicates a velocity of the order of 9 km/sec. As might be expected it is much higher than for the motion of matter near the head.

Objective prism spectrogram W 232, June 1.749, shows features in the CO^+ images similar to those of the direct photograph WF 37, June 1.748. Strong sharp edges of Tail I up to 40′ from the nucleus can be better seen on the violet side of the CO^+ images. This edge is brighter on the direct photograph and the CO^+ bands degrade to the red. At a distance of 40′ from the nucleus a rather sharp bend is well seen in the CO^+ images. A condensation in Tail I, 2°45′ from the head, and another condensation at 4°50′, are visible both on the direct photographs and in the CO^+ images.

The CN IV knot has a fuzzy halo of 22′ diameter which corresponds to the outermost halo visible only on the overexposed Crossley plates. The CN IV and

C IV knots have a bright halo of 8′ diameter corresponding apparently to the halo of 13′ measured on the direct photographs. Measurement in both cases is uncertain due to the fuzziness of the images. This halo is also present in the CN III and C V bands on the sunward side but it is very faint. Estimated intensities of the 8′ ahlo: C IV, 10; CN IV, 9; CN III, 2; C V, 1. The core of the spectral images is 4′ in diameter corresponding to the innermost bright halo. The nuclear line on W 232 is from 10″ to 15″ in width.

The nucleus appears to have swollen within the short interval covered by the direct photographs. Its diameter on C 2298 is 5.″8; on C 2301, at least 8″, but precise measurement is impossible.

Barnard, on June 1.590, describes the nucleus as very faint and small, just visible in a strong condensation (apparently inner halo). By June 1.624 it could no longer be seen to guide on.

Borrely [66] at Marseille on June 1.375, observed a very brillant jet 1′ long directed toward the Sun. He described it as very similar to that of May 31 and his inference was that the jet of May 31 had rotated 180°. This idea was strengthened by the appearence of a jet on June 2 in the direction of the prolonged radius vector. These were entirely different formations, as appears from a detailed investigation. There was no appreciable rotation and the equal length of the jets on May 31 and June 2 was a mere coincidence.

The jet observed by Borrelly on June 1 is apparently jet *h* of this date (see fig. 76c). It is by no means brilliant, it is indeed much fainter than the rest of the nuclear jets on that date. H. C. Wilson[67], on June 1.649, describes jets *c* and *d* but does not mention *h*. It must have decreased in brightness very considerably between June 1.375 and 1.649.

June 02; observations see Fig. 8

On this date another remarkable outburst occurred in the nucleus and adjacent portions of the head.

Innes, at Johannesburg (June 2.208), notes: "Nucleus is broken up. The chief nucleus is probably the preceding one, which is slightly the brightes. It is followed in p.a. 120°, 20″ ± by a double nucleus, also double a long 120°, 5″ ±, and a band of light joins all three nuclei." On the original drawing on June 2.213, another very faint and small nucleus is shown approximately in p.a. 180° from the principal nucleus. The three bright nuclei are of the same brightness. On June 2.239 (44^m later), the preceding nucleus was very much brighter than the others and the fourth faint nucleus changed its place so that it was nearer the line joining the three brighter nuclei. The p.a. of the double nucleus in respect to the principal one on June 2.281, was 111.°5.

Worsell also noticed the change in the position of the double nucleus with respect to the principal one. On June 2.260, the p.a. was 112°, distance 22″. On June 2.339, p.a. 115°, distance 27″.

Javelle[69] (Nice) measured the p.a. of a jet as 110° on June 2.357. This jet was apparently the same formation as the brighter secondary nuclei described by the Johannesburg observers. Borrelly [70], June 2.366, observed a brilliant jet about 1′ in length, directed along the axis of the tail. Its tip appeared like a secondary nucleus 41″ distant distant from the principal one.

Quénisset[71], in his drawing made at June 2.386, gives the jet in p.a. 112° with several bright points in its tip.

Giacobini[72], on June 2.417, describes the comet as clearly double. It is possible to distinguish two components, each surrounded by a nebulosity spherical in shape. In their centers are well defined nuclei. The fainter one is of the 10th magnitude. The differential measure gives $\Delta\alpha = 1^s.64$, $\Delta\delta = 8''.1$ (which corresponds to p.a. 108°, $d = 26''$).

Rambout[73] at the same time, June 2.417, says: "the horn extending from the nucleus in p.a. 120° is more marked than on May 31."

From these and similar descriptions it is obvious that there was a good deal of confusion among observers as to the significance of details they were measuring. It seems probable that besides the bright jet in p.a. near 111° with several condensations in it there was also a condensation like a secondary nucleus in p.a. 120°.

On the Mount Wilson plate MW 22, June 2.685, there are clearly two nuclei. The principal nucleus n_1 is much overexposed and its diameter is 10″.8. The almost simultaneous Crossley photograph C 2303 (June 2.688), shows that true nucleus is only 3″.9 in diameter. It is on the periphery (toward the Sun) of the dense spherical envelope 10″.8 in diameter.

The secondary nucleus n_2 on MW 22 is in p.a. 123° and 16″ distant from the principal nucleus. It is 5″ in diameter and is very much fainter than the principal one. Both nuclei, n_1 and n_2, have curved jets *mn* and *pq* which can be traced to the boundary of halo α. (See figure 77) Both nuclei and the straight jet are connected by a thread.

The straight jet on MW 22 is in p.a. 109° (axis), it is 1′ 48 long, and its maximum width is 24″. There are four spherical halos α, β, γ, δ of which α is displaced into the tail with reference to the nucleus by 0′.3, β by 0′.6. Halos γ and δ have the primary nucleus in their center.

This photographs is reproduced in Plate 18, figure 44. All the details described above can be identified with the help of the drawing of figure 45, Plate 18 (C 2305, June 2.733). On this reproduction some of the details of figure 44 can be made out. The halos on figure 44 are not quite homogeneous. They are clearly separated by shells of darker matter; this shows that they were formed by explosions of short duration.

MW 23, June 2.729, shows some changes in comparision with MW 22. The straight jet has increased in size to 1′58″; it now protrudes out of halo α. Its p.a. is 111°. The secondary nucleus is in p.a. 123°, 24″ from the primary. There seem to be changes also in the relative intensity of the halos.

On the Helwan plate of June 2.299, the halos are easily recognizable. The results of measurement are given in the following table:

Table 93

Halo	Diameter			Velocity of Expansion	
	(1) Helwan	(2) Mount Wilson	(3) Mount Wilson		
Date	June 2.299	2.685	2.729	(1)—(2)	(2)—(3)
Nuclear		10.″8	16.″2		0.29 km/sec
α	1.′5	2.′4	2.′4	0.33 km/sec	0.00
β	4.1	4.8	5.4	0.25	1.96
γ	5.0	7.2	.7.4	0.84	
δ	15.0	15.3			

There seems to be no doubt of the acceleration of the expansion of halo β. This gives evidence of the existence of a repulsive force originating in the nucleus.

Objective prism spectrogram W 234, June 1.748, shows CN IV, surronded by four successive envelopes corresponding to halos α, β, γ, δ. Their diameters were measured as follows: 2.′9, 6.′5, 10.′2, 16.′5. The C IV band has an inner bright envelope of 8.′1, apparently β and γ. Envelope δ is completely absent although the band itself is brighter than CN IV. All other radiations CN III, C V, C + H are restricted to halos α and β.

There are jets about 3′ in length in the CN IV and C IV bands which might be due to the straight jet *s* lengthened by guiding. No such jet can be seen in the CO^+ images. These latter are entirely distinct from the halos.

The CO^+ images have some details identified with similar details on the direct photograph WF 38, June 2.747. Such are: bifurcation of Tail I at 36′ from the nucleus, condensation at 2°48′ in the soth branch of Tail I, and some other features. The bright jet, or else the beginning of streamers (see Pl. 18, fig. 45, central part) of 11′ length is well visible in C IV, C V, CN IV, CN III, and is especially strong in C IV. Nothing like this is observed in the CO^+ bands. Here again is an illustration of the fact that the fine streamers have a different spectrum (CO^+) from that of their parent jets. Some streamers near the head of the comet can be traced to their source. They are denoted in figure 45, Plate 18, with the letters corresponding to those of the jets in figure 77 [see Fig. 8].

... The conditions of ejection can be approximately calculated. If we assume that the ejection occured on June 2.272, we have $g = 0.21$ km/sec; $G = 46°$; on

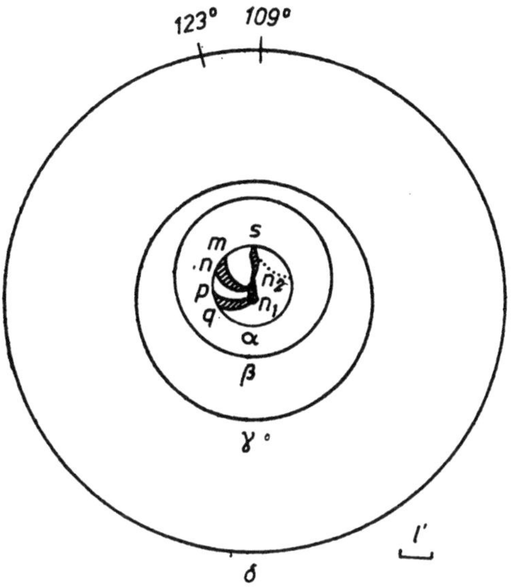

Fig. 8. The structure of the head on June 2.685; N. T. Bobrovnikoff, Publ. of the Lick Observatory, Vol. XVII, pt. 2, 1931, p. 405

the assumption of the moment of ejection as June 2.200, $g = 0.35$ km/sec, $G = 25°$. The latter value is probably nearer the truth. In both cases the velocity of ejection g is very small, of the same order as that derived from the expansion of the halos on this date. The direction of ejection can not be determined exactly ...

It must have decreased very rapidly, approaching within a few hours the p.a. of the prolonged radius vector (112°). Some confirmation of this may be found in the measurement by Innes discussed previously. The p.a. of the secondary double nucleus on June 2.208, was 120°; on June 2.281, only 111°.5. It is therefore probable that the secondary nuclei of Innes were but condensations in the jet. Indeed, the jet was described as a secondary nucleus by many European observers.

[66] Bull. Astr. **27**, 365, 1910.
[67] Pop. Astr. **18**, 477, 1910.
[68] Compt. Rend. **151**, 44, 1910.
[69] Bull. Astr. **27**, 456, 1910.
[70] Bull. Astr. **27**, 365, 1910.
[71] Bull. Soc. Astr. France **24**, 367, 1910.
[72] Bull. Astr. **27**, 394, 1910.
[73] Radcliffe Obs. Monthly Notices, **71**, 139, 1910.

N. T. Bobrovnikoff, Publ. of the Lick Observ., Vol. XVII, pt. 2, p. 402—407, 1931.

Juni 03; observations see Fig. 9

MW 25, June 3.725, shows a fuzzy and indistinct nucleus of 6″ diameter (C 2307). It seends out a very strong curved jet *pq* (fig. 78 a) with edges in p.a. 214° and 248°. It is apperently the same jet as on June 2 (fig. 77) but is much stronger in comparison with other parts of the head. The straight jet *s* is considerably increased in size and its tip is hardly distinguishable from the rest of the streamers. the secondary nucleus n_2 of June 2 has almost entirely disappeared. Instead of it there is a jet, n_2, of 1′ length in p.a. 121°. Jet *mn* of June 2 is not discernible. There is a peculiar formation *t* in the third quadrant which resembles a globe of 54″ diameter on the periphery of which is the primary nucleus n_1. The center of the globe is situated 27″ from the primary nucleus, in p.a. 318°. This center has no condensation or anything of the nature of a secondary nucleus but a strong curved jet *r* issue from it in p.a. 243° in reference to the center of the globe. This makes it probable that there is in reality a secondary nucleus concealed by the gaseous envelope.

It is possible to trace the origin of the streamers in the head to their jets. The double streamer *PQ* (PL. 19, fig. 46) is due to jet *pq*. Jet *r* gives rise to the lateral faint streamer *R*. A number of streamers *S* issue from the central jet *s*. The streamers on the north side of *S* begin somewere in region *t* but cannot be traced to their source.

The other MW 24 plate does not show any change in comparision with MW 25. The four Crossley plates 2307—10, of different exposure times, bring out various details. Without them the identification of the structure on MW 25 would have been impossible. The difference in brightness of the details in the head is so great that only a series of photographs with widely different exposure time can give an adequate idea of the structure.

On comparision of these plates with Helwan H 45, June 3.274, it appears that jet *r* developed in the interval between June 3.3 and 3.7. Streamers *PQ* and jets *pq* are well visible on H 45.

There are at least three halos on MW 25 (PL. 19, fig. 46). Their diameters and brightness are: α, 1′.8, β, 4′.8, 5; γ, 10′.2, 1. The center of halo β is displaced 1′.0 into the tail.

Plate WF 39, June 3.748, shows that the lateral streamer *R* extends but 35′ from the nucleus. Tail I is formed chiefly by streamers *P* and *Q*. On Z 18, June 3.748, Tail I can be traced to 25° from the head. The diffuse Tail II is also well visible.

Objective prism spectrogram W 235, June 3.748, has very distinct images of CO^+. A bend in streamers *P* and *Q* at 25′ from the nucleus is well shown. There is also a fork 2°.4 from the head on the direct photographs in Tail I and in the CO^+ images. The C + H, CN III, C V images have halos of 4′.7 diameter, apparently halo β. The bright knots of CN IV and C IV have diameters of 13′, corresponding to halo γ. The cores of C + H, CH III, C V have diameters of 1′.4, corresponding to α. There is a fuzzy halo extending about 20′ into the tail in CN IV (but not in C IV), which corresponds to the region where the streamers are formed. It cannot be due to the streamers themselves because the sharp edge of *P* and *Q*, well visible in the CO^+ images, is absent in the halo. Something of the nature of this larger halo is visible in C 2309.

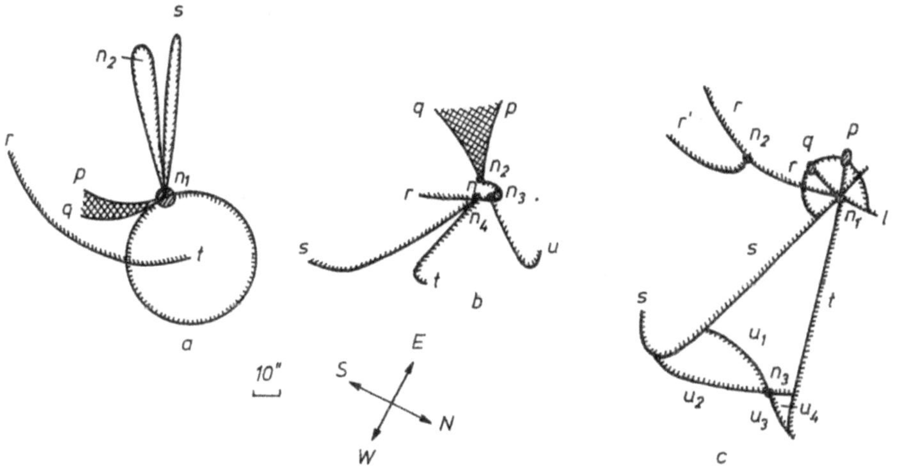

Fig. 9. The head of the comet on (a) June 3.725, (b) June 4.694, (c) June 5.679, N. T. Bobrovnikoff, Publ. of the Lick Observatory, Vol. XVII, pt. 2, p. 408

At Johannesburg the nucleus was observed like a double star with p.a. 200°, distance 1″. This direction corresponds to jets *pq*. The difference in magnitude between the two nuclei was 1ᵐ0; the color was blue-yellowish, probably due to C_2 (Swan spectrum). The double star appearance persisted from June 3.198 to June 3.288. It is possible that the escaping gases which later formed jet *pq* pushed a part of the nucleus toward p.a. 200°. Later it again joined the primary nucleus.

N. T. Bobrovnikoff, Publ. of the Lick Observ., Vol. pt. 2, S. 407/408, 1931.

Juni 04

El núcleo del cometa de Halley el 4 de junio

Como complemento necesario á mis anteriores observaciones del núcleo del cometa de Halley, me apresuro á remitir á las A.N. las que acabo de efectuar el 4 de junio, y que revisten excepcional interés.

El núcleo secundario señalado el dia 2 de junio ha sido invisible el dia 4. En cambio, aparece un agrupamiento sorprendente de nuevos núcleos secundarios ó condensaciones. Aparte de *a*, que es propiamente el núcleo del cometa, se distinguen, por lo menos, cuatro núcleos ó condensaciones más. Para no prejuzgar la naturaleza de estas apariciones, las ilamaré simplemente globos. El globo mayor es *b*, cuya magnitud total es una unidad y media inferior á la del núcleo *a*. Los demás globos son bastante pálidos, pero fáciles, exceptuando el señalado con (?), que es dudoso.

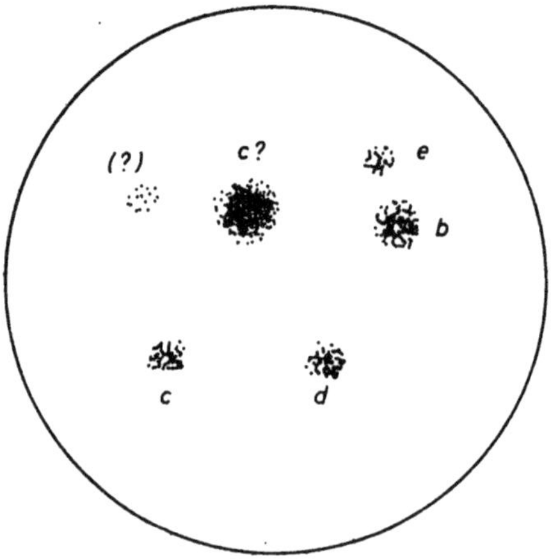

Fig. 10. Multiple nuclei of comet Halley, observed by J. Comas Sola, Astron. Nachr., Nr. 4421, 1910, S. 76/77

El 4 de junio, empecé la observacion á las 8^h40^m, y al cabo de un cuarto de hora era ya sensible que los globos se separaban radialmente de a. Para determinar este movimiento, realié una serie de medidas micrométricas sobre los puntos *a* y *b* (hilos de platino). He aqui los resultados:

4 junio 1910

Horas	Distancias	Horas	Distancias
8^h45^m	6.''5	10^h5^m	11.''8
8 50	7.0	10 20	12.8
9 40	7.7	10 35	12.4
9 50	9.9		

En cuanto al ángulo de posición, no ha variado sensiblemente de un mismo valor = 123°23′, que corresponde al de la dirección de la cola, en el sentido opuesto al Sol.

Combinando las anteriores observaciones de distancia de tal manera que intervengan todas ellas en el cálculo, resulta que la velocidad media de separación del globo *b* al núcleo *a*, desde las 8^h45^m hasta las 10^h35^m, fué de 4.″28 por hora, lo que da una componente transversal de la velocidad propia de dicho globo de unos 500 metros por segundo (más exactamente, 527 m.). El globo *b* debió, pues, surgir del núcleo *a* hacia las 7^h32^m del propio dia.

Estas observaciones demuestran que la bocanada que observé el 31 de mayo no fue el núcleo secundario ó globo del dia 2 de junio, sino que fueron objetos distintos. En fin, hasta ahora, parece que estos globos se disipan ó se hacen invisibles en cuanto llegan á una determinanda y no muy grande distancia del núcleo principal (unos 25000 kms). Ecuatorial doble de 38 cm; aumento, 250 y 150 veces.

Observatorio Fabra, Barcelona, 5 junio 1910.

Astr. Nachr., Nr. 4421, S. 76/78, 1910.

MW 26, June 4.694, shows remarkable details (fig. 78 b). The nucleus is imbedded in a triangular formation n_2 n_3 n_4 in which can be dicerned at least four condensations. The central one, n_1, is the brightest of all and is here called the primary nucleus. It is not perfectly spherical but its average diameter is 3″. The position-angles, diameters, and the distances of the centers of the secondary nuclei from the center of the primary are given in Table 98.

Table 98

Nucleus	Diameter	Position angle	Distance
n_2	2″	127°	5.″5
n_3	5	19	4.0
n_4	2	232	2.5

The space between the nuclei is filled with luminous matter which may contain other centers of condensation, especially in p.a. 70°.

The secondary nucleus n_2 has two bright, slightly curved jets *p* and *q* (the notation here has nothing to do with that of June 3). These jets issue in p.a. 102° and 145° counting from n_2. Jet *p* is almost on the prolonged radius vector (p.a. 112°) and its motion was stidied in detail. The secondary nucleus n_4 has a short slightly curved jet *r* of about 22″ length, p.a. 205° and two long jets *s* and *t*. Jet *s* issue from n_4 in p.a. 237° and reaches the vertex 62″ from the primary nucleus in p.a. 232°. Jet *t* issues in the same p.a. but curves in the opposite direction reaching its vertex 40″ from the primary, in p.a. 252°. There is one more jet *u* issuing apparently from the primary nucleus in p.a. 314°. Its vertex is situated 32″ from the primary in p.a. 324°.

The portion of the tail adjacent to the head is full of fine distorted streamers (nearly lost on the reproduction, Pl. 19, fig. 47). It is difficult to trace them to their source because of the complexity of the structure.

Some of the condensations in the streamers could be measured on successive Mount Wilson plates. They show rapid motion with a very large acceleration. Three streamers near the prolonged radius vector were selected. Their p.a.'s are as follows: A, 90°, B, 127°; C, 106°, counting from the primary nucleus ...

Assuming that the motion was in the radius vector we obtain $(1 - \mu) = 488$ for A; 848 for B; 878 for C. The cometocentric velocities for the same date are A. 20 km/sec; B, 14 km/sec; C, 16 km/sec. Thus *A* was ejected with a much higher velocity than *B* and *C* but was subject to a smaller repulsive force.

The values of the repulsive forces are of course only approximate but there is no doubt that they were very high in contrast with the jets and other formations near the head.

Plate C 2313, June 4.741, shows especially distinctly three halos (also visible on the reproduction Pl. 19, fig. 47). The inner one, α, 1′.4 in diameter, is broken up in the third quadrant; the middle one, β, 5′.6, bright and sharp toward the Sun, diffuse toward the tail, is displaced 1′.5 along the axis of the tail; the outer halo, γ, is 14′ in diameter.

Objective prism spectrogram W 236, June 4.739, shows sharp CO^+ images in which the following details can be distinguished, corresponding to those on the direct photographs; two bright tail streamers, the central one and the northern, lateral up to 22′ from the nucleus; a sharp bend to the south at 1°33′ from the head; a fork at 2°50′. The diffuse Tail II at C IV does not show these details; its axis makes an angle of 5° with Tail I both on the spectrogram and on the direct photograph. CN IV has a diameter of 19′ (corresponds to halo γ). The inner bright portions of CN IV and C IV are 5′.4 in diameter (halo β). CN III, CV, C + H are 1′.2 in diameter (halo α).

Helwan plate, HR 48, of June 4.329, shows a strong curved jet between p.a. 191° and 202°. This is probably jet *pq* of June 3 (p.a. 214°—248°). If this is so, then jet *rs* of June 4 (p.a. 205°—237°) is identical with *pq* of the preceding dates. It is difficult to say whether there was real oscillation of the jet. The nucleus on the Helwan plate is of the peculiar triangular shape observed later on MW plates.

Comas Solá[75], at Barcelona on June 4, saw four or five secondary nuclei the most prominent of which, *b*, was in p.a. 123°. It evidently corresponds to n_2 (p.a. 127°) of the MW plates. From his drawing it follows that his secondary nucleus *c*, p.a. 10°, is n_3 (p.a. 19°) of fig. 78b; his nucleus *d* in p.a. 66° may also be suspected on the MW plates. Comas Solá's nucleus *e* in p.a. 150° corresponds probably to jet *q* of figure 76b.

This configuration probably developed suddenly. Rheden at Sonnwendstein[76] (time not given, apparently shortly before the observation by Comas Solá) did not observe any especial activity.

At Johannesburg, on June 4.271, the nucleus was observed larger than on the night before. It was of yellow color and looked like a double star out of focus in p.a. 140°; the distance between the components was 4″ ±. Apperently this was the secondary nucleus n_2.

The persistence of the secondary nuclei is the more remarkable since they were in a state of high activity. Comas Solá observed a continuous motion of formations which he calls "globes". He measured seven positions of those moving along the line $n_1 n_2$. The displacement in 1^h50^m was 5.″9. This gives the average velocity 0.53 km/sec. S. V. Orlow[77] made an attempt to compute $(1 - \mu)$ from these data. The value of $(1 - \mu)$ for the whole series was 15, but if the last observation is neglected $(1 - \mu) = 54$. This is of course only a rough estimate.

It may be said here that the velocity of recession of the globes of Comas Solá is of the same order as the velocity of expansion of the halos. The globes also expanded and disappeared on the background of the head.

June 05

The structure of the head is exceedingly complicated. On MW 30, June 5.689, the nucleus is small, about 3″ in diameter. At least five straight jets can be recognized whosep. a. and length are as follows: *l* 0°, 15″; *m* 75°, 15″; *p* 105°, 19″; *q* 159°, 20″; *t* 280°, 94″ (see fig. 78c). Jets *p* and *q* have near their tips condensations-rembling secondary nuclei. A long jet 2′ 55″ in length can be traced in the same p.a. as jet *p* (see Pl. 19, fig. 48). It is probably the remnant of jet *p* of the preceding night. A sharply curved jet *r* issues from the nucleus n in p.a. 200°. At about 40″ from the primary nucleus n_1 it forms a secondary center of emission, n_2, with the curved jet *r′*. The original jet *r* continues without any interference on the part of n_2. Another jet *s* issues from n_1 in p.a. 250°. It is straight for about 80″, then suddenly curves and reaches its vertex in p.a. 245°, 94″ from the nucleus. The space between jets *s* and *t* is full of complicated structure in which can be distinguished a secondary center of activity n_3 in p.a. 274°, 80″ from the nucleus. This secondary center forms four jets *u*, which coalesce with the long primary jets *s* and *t*. It seems probable that jet *p*, *q*, *r* are the jets denoted by the same letters on June 4. Jets *s* and *t* of June 4 may be the same jets of June 5. Jet *m* was suspected on June 4 while *l* is an entirely new formation. The secondary nuclei n_2 and n_3 are difficult to identify with any nuclei of June 4. On Helwan plate, June 5.306, traces of n_3 of June 5 are already noticeable, but jet *u* of June 4 is gone.

Streamers P_1 (broad) and P_2 (narrow) show very rapid motion. Unfortunately it was not possible to measure it with any degree of precision owing to a rapid deformation of the streamers. Figure 49 represents C 2317, June 5.736. The axis of P_1

is in p.a. 110° (prolonged radius vector, 113°), that of P_2, in p.a. 123°. The two streamers are not uniform in brightness and show traces of a wavy structure. The first Helwan plate HR 49, June 5.267, shows very well that streamers P_1, P_2 issue not from the nucleus but from straight jet *p* of June 4. The curved jet *r* of June 4 extends about 1′ almost intersecting P_2; it is clear, however, that it has nothing to do with P_2. There are on the Helwan plates (June 5.306 and 5.283, HR 50) two bright lateral streamers which are almost entirely absent on the MW plates. These lateral streamers intersect the bright envelope without the slightest disturbance.

Objective prism spectrogram W 237, June 5.748, shows the nucleus line 1.′6 in width, corresponding to the innermost bright envelope.

A halo around CN IV and C IV (much fainter) corresponds to the outer spherical halo of 13.′5 diameter. The branching of streamer P_1 (forming Tail I) at 3° from the nucleus is well seen in the CO^+ images.

[75] Astr. Nachr., **185**, 75, 1910 (drawing).
[76] Astr. Nachr., **185**, 243, 1910.
[77] Russ. Astr. J. **2**, III, 14, 1925.

N. T. Bobrovnikoff, Publ. of the Lick Observ., Vol. X VII, pt. 2, p. 408—411.

June 06

Eine Photographie zeigt den Schweif in 3 Spitzen gegabelt.

Astr. Nachr., Nr. 4431, S. 243, 1910.

On the Acceleration of the Receding Masses in the Tail of Halley's Comet 1910 June 6

By *E. E. Barnard*

The opportunities here for determining the motion of the tailforming particles of Halley's comet have been few, definite masses suitable for the purpose were seldom shown on the plates. Cloudy and smoky skies almost wholly prevented satisfactory morning observations while the comet was at its brightest. When it appeared in the evening sky the conditions were better, and the comet was a splendid object. The photographs of May 26 showed a peculiar twisted appearence in the tail, which was moving rapidly away from the comet. But the most interesting phenomenon occured in the pictures of June 6, when a large separate section of the tail was shown, the rear end of which was about 1.°5 from the head. The successive positions of this can be seen, for instance, in photographs taken by Morehouse at des Moines, Iowa Ellerman at Honolulu, and by Joy at Beirut, Syria. These photographs, or copies of them, have been placed at my disposal by these gentlemen, and I have measured the position of the rear end of the receding tail on the various pictures, and have deduced from these measures the motion of the particles of the tail.

The interval between Mr. Morehouse's photograph and mine, 22 minutes, is too small to be used in the investigation, but the picture serves admirably as a check. The intervals of time between the Yerkes Observatory photograph and those of Mr. Ellermann and Mr. Joy are 4^h15^m and 15^h9^m respectively. These intervals are sufficiently large to show the change in the motion of the mass.

Because of the diffused nature of the object, it was not possible to make the measures on a measuring machine. They were therefore made with dividers and a scale of inches. This method seems accurate enough for the purpose.

The intention of this paper is to give only the results of this investigation. The details will be printed in a later paper on the comet. It is interesting to state, however, that the distance of the rear end of the tail from the head of the comet at the times of the different photographs was 2007000, 2364000, and 3847000 miles, respectively.

The following table will show the results that have been obtained from the investigation of the three photographs.

		Recession per second				
		Hourly	From Comet		From Sun	
Station	Interval	Motion	Miles	km	Miles	km
Y, O — Honolulu	$4\overset{h}{.}25$	3:60	23.1	37.2	39.7	63.9
Y.O. — Beirut	15.15	5.17	33.1	53.3	49.7	80.0
Honolulu — Beirut	10.90	5.78	37.3	59.7	53.9	86.4

At this time the recession of the comet's head from the Sun was 16.6 miles (26.7 km) per second.

These results show a strong acceleration in the receding mass. The acceleration in the last two photographs was about 14 miles ore 22 km per second.

Many interesting changes in the form of the disconnected portion of the tail took place between the photographs.

E. E. Barnard

Yerkes Observatory, Williams Bay, Wisc.

Astr. Nachr., Nr. 4441, S. 11/12, 1911.

Velocities of Particles in the Tail of Halley's comet

By *Percival Lowell*

The photographs taken here of Halley's comet have furnished experimental evidence of the mode of genesis of the tail. Measurement of certain irregularities in that appendage has shown its particles to moving under the action of a repulsive force directed from the Sun.

Although the comparatively orderly behaviour of Halley's comet has made measurable caudal irregularity rare, it has been possible on occasion to detect lapses from monotonous uniformity sufficienty pronounced to be capable of identification on successive plates. On May 23, 1910, two such plates were taken by Mr. Lampland, Dr. V. M. Slipher and Mr. E. C. Slipher with a Brashear doublet of five inches aperature, one exposed from 9^h23^m to 9^h58^m; the other from 10^h to 10^h53^m. The irregularities are of two kinds, thickenings or thinnings of the tail in places and aprupt changes of direction in it; knots and notches. As these are of necessity long drawn out great accuracy is not possible and therefore the resulting determinations for the velocities cannot be considered exact. The intresting thing about them is their pointing unmistakably to solar repulsion ...

We have for the distances in degree from the head of the several knots examined

	Apparent	Corrected for refraction	Corrected for inclination
Knot 1	43.5	1°24′	1°28′
2	96.3	3 6	3 12
3	139.0	4 28	4 36
4	189.8	6 6	6 15

and for the differenciated distances corrected for inclination:

Knot 1	6′.5	Knot 3	9′.4
2	8.2	4	14.2

giving for the speed of the particles composing the knots:

Knot 1	13.6 miles a second	Knot 3	19.7 miles a second
2	17.2 miles a second	4	29.7 miles a second

Here we are confronted by an increase in velocity away from the head as the particle gets farther away, an exes of speed which betrays the action of a repulsive force. Owing to the diffuse character of the knots more detailed information is not possible.

Percival Lowell

Astr. Nachr., Nr. 4441, S. 13/14, 1911.

Nota sobre las velocidades y aceleraciones de una bocanada del cometa de Halley
Por *José Comas Solá*

Unas fotografias del cometa de Halley obtenidas en el Observatorio de Madrid, y que ha tenido la amabilidad re remitirme su director don F. Iñiguez, me Ilevan á completar las observaciones mias que se publicaron en las A.N., no. 4421, referentes á la emisión, del nucleo del cometa, de un grupo de globos fosforescentes, el 4 de junio.

Del examen de una fotografia del señor Iñiguez y de otra obtenida por mi 80 minutos antes (momentos medios de las exposiciones) resulta que en ambas fotografias aparece una importante bocanada gaseosa, con la circunstancia de que el movimiento propio de la bocanada, con relación al núcleo, se hace perfectamente sensible en este intervalo de 80 minutos (desde las 21^h21^m (tiempo civil de Gr.), que corresponde á mi fotografia del señor Iñiguez). Midiendo este desplazamiento sobre las fotografias, resulta para la bocanada una velocidad media igual á 37425 metros por segundo. En estos cálculos, considero que la perpendicular á la dirección la cola trazada desde la Tierra formaba, aquellos dias, un ángulo de 30° con la visual dirigida al núcleo. Supongo tambien que, durante estos mismos dias, la paralaje media del cometa fuí de 12.″5, y que la trayectoria de las proyecciones era rectilinea.

Debo recordar ahora que el globo más brillante, cuyo movimiento pude medir el dia 4 de junio, hacia las 22^h, llevaba una velocidad media transversal de 527 metros por segundo. No creo estar muy lejos de la verdad admitiendo que la velocidad real de dicho globo (y probablemente de la de los demás) fuese, en números redondos, de 600 metros por segundo. Esto supuesto, podemos tomar ambas velocidades medias como expresión muy aproximada de la velocidad instantánea de la bocanada y de los globos en cada una de dichas posiciones. Además, contando con otra fotografia obtenida por el señor Iñiguez el 7 de junio, en la que aparece también dicha bocanada, aun cuando mucho más alejada del núcleo, será posible calcular las aceleraciones medias de la misma á diferentes distancias del núcleo.

En todas estas consideraciones doy como cierto, y asi todos los hechos tienden á confirmarlo, que los globos que descubri el 4 de junio han dado origen, expansionádose, á dicha bocanada, proceso de formación que, desde luego, ya consideré general en todos los cometas, conforme hice notar en otro trabajo ("Comptes rendus de l'Académie des Sciences de Paris", sesión del 20 de junio de 1910).

He aqui una pequeña tabla de datos y de resultados:

Fecha	Horas	Distancia al núcleo en kms	Velocidadas instantáneas en metros	
			Medidas	Calculadas
1910 junio 4	21^h40^m (globo)	4536	600	—
6	22 1 (boconada)	2356584	37425	26426
7	21 45 (id.)	6457736	—	58575

Para calcular las aceleraciones medias entre el 4 y el 6 de junio y entre el 6 y el 7 de junio, de la conocida fórmula $e = {}^1/_2\alpha t^2$, y llamando V_m á la velocidad media en un intervalo de tiempo, descontada la velocidad adquirida en el otro intervalo, se obtiene:

$$\alpha = \frac{2V_m}{t}.$$

He aqui las aceleraciones medias que resultan:

Intervalos de tiempo	Aceleraciones medias en metros por segundo
Del 4 al 6 de junio	0.148
Del 6 al 7 de junio	0.248

Como se ve, existe una diferencia notable de aceleracion entre uno y otro periodo ó intervalo, diferencia que viene confirmada, según aparece en la primera tabla, por un notable aumento de la velocidad instantánea, media, el 6 de junio, sobre las fotografias, respecto de la que resulta de la velocidad media correspondiente al intervalo del 4 al 6 de junio.

El aumento de aceleración pudiera explicarse, cuando menos, por las dos siguientes hipótesis: 1ª Admitiendo una fuerza atractiva en el núcleo del cometa; 2ª Tentiendo en cuenta el aumento de superficie expuesta al soplo solar, al dilatarse enormemente los globos para transformarse en una inmensa nube ó bocanada.

La primera hipótesis es admisible, pero es poco probable, tratándose de distancias considerables al núcleo; la segunda puede satisfacer completamente, considerando que, á pesar del extremo enrarecimiento de las proyecciones cometarias, puede aumentar por modo muy sensible la acción de la fuerza repulsiva del Sol al pasar estas masas gaseosas desde el pequeño volumen de los globos del 4 de junio á la colosal magnitud de bocanadas como la fotografiada los dias 6 y 7 de junio. Es evidente que esta circunstancia imprevista tiene que dificultar muchisimo la determinación de la fuerza repulsiva del Sol, en un momento dado, sobre moléculas gaseosas supuestas de una misma naturaleza.

En cuanto á las variaciones de la fuerza atractiva del Sol, no han influido ciertamente de un modo sensible en toda esta serie de fenómenos observados.

Astr. Nachr., Nr. 4441, S. 13—16, 1911.

A long series of photographs collected by Dr. Curtis from various observatories leaves no doubt that the remarkable condensations of June 6 developed out of jet *p* of June 4. Plate 19, figures 49 to 51 and Plate 20, figures 52 to 58 represent the development of this condensation. It evolved wholly out of streamers P_1. The other streamer P_2 dissipated very rapidly and had already disappeared on June 6.498 (Cordoba). The condensation in streamer P_2 is barely visible on June 5.736 (Pl. 19, fig. 49) and June 5.746 (Pl. 19, fig. 50). Its traces (near the junction of $P_1 P_2$) could be seen as early as June 5.489 (Cordoba). It is already fully developed on the Dairen Plate of June 6.068 (Pl. 19, fig. 51 and Pl. 20, fig. 53). On June 6.498 (Pl. 20, fig. 54), Cordoba, it clearly begins to differentiate into a southern sharp streamer *A* and northern diffuse streamer *E*. The northern streamer begins to break up into a large number of condensations. Further plates show the continuation of this process which suggests selective repulsive forces in the tail.

It seems safe to conclude that the condensations in the northern and the southern branch moved under quite different repulsive forces, a fact which may be attributed to their different constitution. Objective prism spectrogram W 238, June 6.745, corroborates this surmise. It is reproduced in Plate 20, figure 59, and may be compared with the simultaneous direct photograph WF 42 reproduced in figure 55 of the same plate. It will be seen that the CO^+ images in general repeat the image of the tail with, however, significant differences in details. While on the direct photograph *A* and *E* are approximately the same brightness, on the spectrogram *A* is very much brighter than *E*. On the other hand the diffuse image of the tail near C IV does not show the characteristic curvature of Tail I (that is, of streamer *A*). In the images of CO^+ at λ 4000 and λ 4251 not only are the details of the southern streamer visible but the edges of the envelope forming Tail II are also well distinguishable. The rift between *A* and *E* is striking in these images of CO^+, but *A* is not isolated from the nucleus: there is a strip corresponding to the southern edge of Tail II. But while on the direct photographs this edge is much stronger than streamer *A*, it is quite the reserve with the CO^+ but in different proportions.

The four Crossley plates of June 6 show that the activity of the comet has sub-

sided. There is a fine envelope on the north side. There are jets 30″ long in p.a. 186°, 251°, 5°. The jet in p.a. 251° is the most intense of the three. Emissions are more abundant between p.a. 251° and 5° than between 186° and 251°. It is not possible to trace the origin of the fine streamers in the tail. Motion and change in them is apparent. There inner emission from a semi-spherical envelope α; the distance from the nucleus to the vertex is 0ʹ.5. There is a very bright perfectly spherical halo, β, 4ʹ.3 in diameter, with the nucleus in its center. A fainter outer halo γ, is concentric with β. Its diameter is 6ʹ.5. Still another halo δ is brighter in the direction of the tail; its diameter is 13ʹ.
The nucleus on C 2319 (June 6.689) is about 7″ in diameter.
On the spectrogram already described the inner core of all the spectral images, including CN IV and IV, has a diameter of 1ʹ.6 corresponding to envelope α. middle bright envelope of all the images is 4ʹ.7 in diameter which corresponds to β. The outer envelope of CN IV and C IV has a diameter of 8ʹ.0, that is, of halo γ. The CN IV band extends into the tail 13ʹ, corresponding to halo δ.

Barnard noticed that on June 6.601 the nucleus was star-like, of $8^{m}.5$. At 6.608 the nucleus was "beautifully star — like and imbedded in a very strong condensation that faded rapidly and was itself nebulous". At 6.654 the nucleus was $9^{m}.0$. The comparison star was the same, so that we have to deal here with a real change in brightness.

N. T. Bobrovnikoff, Publ. of the Lick Observ., Vol. XVII, pt. 2, S. 427, 1931.

June 07

... the outer envelope seems to have faded out ... the northern edge of the envelope is much brighter than the southern, just as on the prerding date ...

Tail II ... consists of many fine streamers of which two more prominent. They join condensations A and E to the nucleus. At about 22ʹ from the nucleus they coalesce forming a bright single streamer is well visible in the CO^+ images. traces of condensations A and E can also be distinguished in CO^+.

N. T. Bobrovnikoff, Publ. of the Lick Observ., Vol. XVII, pt. 2, p. 428, 1931.

June 08

June 8. — Nucleus is a stellar point of 7 mag. enenly surrounded by nebulosity.

Transvaal Observ., Circular No. 4, 11th July 1910, p. 30.

June 09

No. 59, June 9. The tail is 5°.5 long and very faint and diffuse exept near the head. An irregular, bright streamer extends from the along the axis of the tail. It divides into two streamers at a distance of 1°.33.

E. E. Barnard, Publ. Astron. and Astrophys. Soc. Am, Vol. 2, p. 224, 1915.

June 10

Crossley plates C 2331, June 10.703, and C 2332, June 10.713, show very strong circular halos: the inner, α, has a diameter of 80″. It is displaced toward the Sun so that the nucleus is almost on its periphery. It is much denser with reference to other parts of the head than on the night before. The outer halo, β, is 125″ in diameter; the nucleus is in its center. The nucleus is diffuse; its diameter may be 15″ to 16″ but precise measurement is impossible. There are traces of a condensation in the tail.

On the Helwan plate of June 10.301 the inner halo is 30″ in diameter. The nucleus is on its periphery opposite to the Sun and is probably identical with halo α of the Crossley plates. If this is so we have for the velocity of expansion of the halo 0.4 km/sec. The center of the halo was moving toward the Sun with the same velocity.

June 11

The three Crossley plates of June 11 show interesting details. On C 2333, June 11.704, the nucleus is diffuse and of irregular shape. Its diameter is about 10″. There is an elliptic appendage *a* about 10″ long in its major axis. It is but slightly fainter than the nucleus itself and may be regarded as a secondary nucleus. There are also serveral other bright formations such as the bright globes *b* and *c*. The diameter of *b* is 40″. The result is that the head appears elongated with major axis 136″ and minor axis 78″ long. The same appearance is preserved on the Johannesburg repruduction of June 12.216.

C 2334, June 11.731, has a halo corresponding in shape to the elliptic configuration near the nucleus. The dimensions of the halo are: major axis 217″, minor 165″. The tail is faint and slender. It consists of a number of seperate streamers. June 11, 1910, was the last date on which the comet could be seen with the naked eye (Barnard).

June 12

Barnard observed (with the 40-inch telescope) on June 12.593, the nucleus ill-defined in a small dense nebulosity 5″ in diameter.

On the Crossley plates C 2336—38 of June 12, globular condensation *b* of the preceding night is much fainter. Its diameter is 70″. If the identification is correct this indicates a velocity of expansion of 15″ per day, or 0.2 km/sec.

There is a faint halo about 4′ in diameter with the nucleus in the center. It is much more intense on the side of the tail. The tail itself consists of separate streamers and is faint. The nucleus is 10″ in diameter (C 2336).

N. T. Bobrovnikoff, Publ. of the Lick Observ., Vol. X VII, pt. 2, S. 429, 1931.

Juin 14. Eclat du noyau 10ᵉ gr.
Juin 23. Nébulosité ronde 2′ environ, éclat du noyau 12^m.

F. Gonnessiat, Astron. Nachr., Nr. 4435, S. 206, 1910.

June 25

June 25. — Not reproduced. — Exposure 5 h. 12 m. to 6 h. 12 m. Short double tail. South branch straight. North branch shorter and curved. Pronomounced break in South tail about 30′ from head. Head very broad, much broader then tail. Several stars visible through it. This Photograph differs from all others in suggesting a granular structure for the head; i.e. it resembles a very close cluster of faint stars. The reasons for this may be photographic.

June 29

June 29. — Not reproduced. — Exposure 4 h. 45 m. to 6 h. 20 m. Faint tail, consisting of a main and almost straight portion, and a narrow branch, wich curves towards the North. The head of the Comet appears simple — a dense with an extended coma.

H. E. Wood

Transvaal Observ., Circular No. 4, 11th July 1910, p. 30.

April 22—June 24

Observations of Halley's Comet at Windsor, N.S. Wales

By *John Tebbutt*

I forward herewith the places of Halley's comet resulting from observations made by me with a square bar micrometer on the 8-inch equartorial. The differential measures have been corrected for errors in the form and orientation of the micrometer, and for the comet's prober motion, but the refraction corrections, being hardly sensible, have been disregarded. The resulting places of the comet are, of course, uncorrected for parallax.

The comet's nucleus was faint in December, and was subject to small fluctuations of brightness. In the morning sky on April 22, civil time, the tail could be traced for two or three degrees from the head, and, owing to the Moon's presence for several mornings, the comet was not a conspicuous object. The Moon having considerably waned, and the sky being beautifully clear. I measured the tail with a sextant on May $1^d17^h24^m$ and found it to be 7°36′ in length. The measures were made by reflecting the head of the comet to the estimated place of the end of the tail. The length on May 7^d17^h, during the absence of the Moon, was 14°10′. The following measures were subsequently made, namely, May $9^d17^h15^m$, 18°7′; May $15^d17^h15^m$ 50°29′; May $16^d17^h30^m$, 57°0′. On the last occasion the twilight slightly interfered, and the tail was observed to deviate slightly to the north of morning of 18th, civil time, was cloudy. The following morning, however, turned out beautifully clear and the Moon had set. The tail at this time was a magnificent object, stretching from the horizon under Venus to a point near θ Aquilae, which was on its northern edge. It seemed to be lost in the Milky Way. It was about two degrees wide at the horizon, and tapered off to a width of about one degree. The head being near the Sun, and, of course, below the horizon, could not be seen. The tail could certainly be traced as far as a point whose R.A. = 20^h10^m, and Declin. = 3° south.

The Sun's disc was closely watched, by projection on a large sheet of white paper, from May $18^d22^h15^m$ to $19^d3^h37^m$, but nothin could be detected beyond the usual solar spots. The atmosphere was remarkably clear till 2^h14^m, when small detached ciouds came up from the south. About the middle of the interval just stated the image of the Sun on the paper appeared to be not quite so bright as it was during the earlier and later part of the examination.

The comet could not again be seen till May $21^d6^h30^m$, when the length of the tail by sectant was 17°18′. On the evening of the 28th the tail was directed to a point two or three degrees south of Jupiter, and its length was measured as 44°13′. On $29^d7^h15^m$, it was seen to better advantage in consequence of the Moon's absence, and its length was 41°49′. On the following evening the estimate was 30°, and on June 5 it was $16^1/_4$°. Cloudy weather then intervend, and on June 12 and 13.2° or 3° only of the tail was visible, owing to the presence of the young Moon. Cloudy weather again prevailed till the 24th, when the comet had pretty well ceased to be visible without optical aid.

During the May observations the comet's nucleus was subject to great variations of brilliancy. An extraordinary increase in brightness was particularly noted on May 30, and the contrast between its appearance then and on the preceding and following evenings was very remarkable.

Monthly Notices of R.A.S. LXXI 3 (1911) 225.

July, 1910

C 2342, July 1.727, shows a diffuse nucleus about 10″ in diameter. A very bright circular halo of 20″ diameter is displaced toward the Sun in such a way that the nucleus is on its periphery. Another fainter halo of 90″ diameter is displaced into the tail so that the distance between its center and the nucleus is 10″. There are two faint streamers forming an angle of 15°. They can be traced through the outer halo to the nucleus. The tail is visible on L17, July 1.727, up to 5° from the head.

C 2343, July 4.723, shows a bright central condensation 12″ in diameter; it seems to be of a triangular shape but this may be due to defective guiding. A faint halo 90″ in diameter and two streamers constitute other features similar to those of July 1. The streamers can again be seen through the halo. There are traces of emission toward the Sun. The tail can be seen to 2°5 from the nucleus (L18).

The central condensation on C 2344, July 5.727, is about 10″, and a faint halo about 40″ in diameter. There are also traces of the tail streamers of July 4. The tail, 2° long (L19), is much narrower than the head giving a typical *Zwiebelgewächs*.

C 2345, July 7.718, shows a central condensation of 7″ to 8″ diameter. There is a fuzzy halo of 90″ diameter. The inner halo about 20″ in diameter is displaced toward the Sun. There are faint ejections toward the Sun and traces of the tail for about 1° (L20).

An interesting observation was made on this date by Eddie of Grahamstown, South Africa (Report): "Head appeared bright with condensed center, constant changes taking place in head. Sometimes it is of great brilliancy throughout, then suddenly shows central condensation."

Last Observations in 1910

The comet was rapidly approaching its second conjunction with the Sun and was lost for serveral months.

On November 11.967, Barnard found it with the 40-inch telescope as round, 0'5 in diameter, very little brighter in the middle. No nucleus could be seen. Its brightness was estimated at 11^m, but the sky was poor.

On December 11.970, Barnard found the comet of 12^m, certainly not fainter than 13^m, round, gradually brighter in the middle. Its diameter was 15″.

C 2427, December 13.027, has a trail of the comet uniformly bright and about 3.″5 in width. C 2428, December 13.069, gives a faint image of approximately the same diameter. The comet is surprisingly small when Barnard's observations are taken into account. On C 2427 one might suspect a faint hazy envelope which would extend for about 15″ but it is exceedingly faint, whereas Barnard speaks of a gradual brightening toward the center. Probably the nucleus did not have visual rays, being composed chiefly of CN and C + H. This would explain the lack of contrast between it and the coma observed visually.

On December 13.979, Barnard measured the diameter of the comet as 15″, its magnitude as about 12. It was not especially brighter in the middle.

C 2131, December 14.020, and C 2432, December 14.036, show the comet as a round nebulosity 4.″5 in diameter. It seems to have a central condensation I″ in diameter.

C 2442, December 30.024, shows a round nebulosity uniformly bright. Its diameter is 4.″6. C 2443, December 30.004, has a trail of the slightest trace of any nebulosity.

N. T. Bobrovnikoff, Publ. of the Lick Observ., Vol. XVII, pt. 2, p. 430, 1931.

Report of the Council to the Comets of 1910

Two comets were visible at the beginning of last year. One of these, Daniell's (1909e), was last seen at Algiers on March 15. but was photographed a month later by Professor Max Wolf at Heidelberg. The calculation of the definitive orbit has been undertaken by Dr. Krassowski, but the period will not differ greatly from 6.4 years. The other comet under observation was Halley's (1909c), which is still visible in large telescopes, though the last published observation is dated December 7. The object is described as of the 14th mag., and the decrease of light will probably be very slow. This comet was observed in every observatory, public and private, and its return has produced and abundant literature; it is impossible to give here any useful analysis ob observations so extensive and varied. This is the less necessary as a monograph is in course of preparation by a committee of American astronomers, which will be the authoritative source for information, aiming, as it does, at the exhibition of the history of the comet's variations from hour to hour. But, apart from the mass of observations, visual, photographic, and spectroscopic, that have been collected, an interest, distinct from that attaching to all bright comets, has centred around a few points to which reference must be made. First, the possibility of seeing the comet was projected on the Sun. It is known that the centre of the comet was projected on the Sun's disc, but, notwithstanding some reported observations, it is doubtful if the comet was seen. It has been suggested that the comet, having a parallax of nearly a minute, would subtend an angle sufficient to include the Sun's disc. Another point which caused great excitement was the possibility of the Earth passing through the tail of the comet. If the tail were simply a prolongation of the radius vector, the question might be easily settled; but since the tail is shown to be of a very complex structure, spreading out as a fan, this question can only be settled by observation. Many obervers have reported unusual atmospheric conditions, but whether such phenomena were traceable to the presence of Halley's Comet is very doubtful. It seems certain that the Earth did not pass through the more conspicuous portion of the tail, and equally probable that it did encounter some of the outlying fan-like extensions. Professor Birkeland suggested the possibility of magnetic disturbances being noticed if, as he supposed, the tail were formed of electrical corpuscular rays. He himself reported a magnetic storm, though no very decided fluctuattions were seen at other observatories. Auroras were seen in several observatories in America, but again there is no proof of any physical connection between these and the comet. Similary with meteors although a stream whose members are seen in the first week of connected with Halley's Comet, no decided proof of connection is forthcoming, nor were more meteors seen than in other years. Many other suggestions were made at the time the comet was increasing in brightness, — evidence of public excitement and of scientific activity, but of little practical service.

Monthly Notices of R.A.S LXXI, 4 (1911) 318

Observations in 1911

On January 3.917, Barnard found the comet round and brighter in the middle. It was about 15″ in diameter and 13^m.

On January 8.896, Barnard estimated the magnitude of the comet to be between $13^1/_2{}^m$ and 14^m; it was gradually condensed toward the middle, but had no definite small condensation. The diameter was 18″ (in Astr. Nach. 187, 189, 1911 the diameter was erroneously given as 33″).

On January 22.849, its diamter was 30″, somewhat brighter in the middle but no definite condensation. On January 29.905, Barnard estimated the comet to be 13^m. On February 7.291, the magnitude of the comet was $13^1/_2{}^m$. On February

21.802, the comet was very gradually brighter in the middle, its magnitude was $14^{m}.5$, diameter 29″.

C 2465, February 22.875, shows a bright uniform condensation $3''.5$ in diameter; also very faint haze about 11″ in diameter in wich the nucleus is eccentrically placed.

On February 26.753, Barnard observed the comet as brighter in the middle with possibly a faint nucleus. Estimated magnitude 13.5 to 14.5; diameter 26″.

On February 28.5, Barnard found the comet brighter than on previous nights. Its magnitude was certainly not less than 13. It was round and brighter in the middle, almost having a nucleus. Its diameter was 30″.

This sudden increase in brightness is well substantiated by other observers. Gonnesiat and Sy[82] (Algiers) found the comet on March 4.399, of 13^{m} which they ascribed to unusually favorable atmosperic conditions. Knox-Shaw[83] found the comet on March 4, at least 1^{m} brighter both visually and photographically than on February 5. The heliocentric distance of the comet at the beginning of March was 4.4 A.U.

Max Wolf[84] (Heidelberg), on March 19.406, estimated the photographic magnitude of the comet 14^{m}. It was 15″ in diameter.

Barnard, on March 19.642, found the comet of $13^{m}.5$, round and a little brighter in the middle. He measured its diameter as 45″, but later thought this was too large. Here again the visual diameter is much greater than the photographic. Both Wolf and Barnard mention that Halley's Comet was much larger and brighter than Faye (1909e), so that difference in measurement can hardly be due to individual judgment.

C 2469, March 25.729, gives the diameter of the comet $3''.1$. Its is uniformly bright

On C 2470, March 27.734, the comet seems to have a central condensation; its diameter is $5''.5$.

C 2471, March 28.728; the diameter of the bright condensation is $3''.5$. There seems to be a faint halo 12″ in diameter in wich the brighter part is imbedded eccentrically.

On April 16.651, Barnard estimated the comet to be of 14^{m} and 30″ in diameter.

On C 2472, April 16.705, the comet is very faint. Its diameter is $5''.0$. On C 2473, April 17.738, its diameter is $4''.7$, C 2477, April 20.742, shows a very faint image. The diameter is estimated at 5″.

Barnard, on April 23.615, found the comet of $14^{m}.5$ to $15^{m}.0$ and 10″ ± in diameter. On April 25.603, he found the comet exceedingly difficult to measure, probably 15^{m} to 16^{m}, and from 5″ to 10″ in diameter.

On C 2488, May 20.744, the comet is exceedingly faint, probably 3″ to 4″ in diameter. On C 2491, May 21.731, its diameter is $5''.8$. Barnard, on May 23.645, found the comet from 15″ to 30″ in diameter with no especial condensation. This was the last visual observation of Halley's Comet.

C 2493, May 27.722, shows only traces of the comet. C 2497, June 15.715, shows a small hazy patch of utmost faintness. This was the last observation bei H. D. Curtis, and seems to be the last observation of Halley's Comet in general.

It follows from the observations late in 1910 and in 1911 that the comet was much larger visually than photographically. From the long series of estimates of its magnitude by Knox-Shaw it follows that it was also brighter visually. It will be remembered that this was found to be the case at the beginning of the observations before perihlion, at least so far as the diameter was concerned.

[82] Astr. Nach. 188, 411, 1911.
[83] Helwan Obs. Bull. No. 2.
[84] Astr. Nach. 187, 431, 1911.

N. T. Bobrovnikoff, Publ. of the Lick Observ., Vol. XVII, pt. 2, p. 429—432, 1931.

SUMMARY

The present investigation of Halley's Comet based on 438 original photographs, 271 reproductions, and 77 spectrograms has brought out the following facts:

1. The motion of matter in the tail showed a wide divergence of repulsive force, varying from 20 to 150, and in some less reliable cases attaining valuces of serveral hundred times gravitation.

2. The motion of particles in the tail consisted of short tracks under varying repulsive forces. The computed hyperbolas were only mean paths.

3. In serveral cases interal transformations occurred in the caudal condensation resulting in complete change of the orbital elements $(1 - \mu)$. These changes were especially striking on May 30 and June 6—7, 1910.

4. The hypothesis that the motion of the tail paticles was in the plane of the orbit of the nucleus was a sufficient approximation to the truth in view of the unavoidable roughness of measurement. The contrary arrived at by some investigators was shown to be based on an erroncous assumption.

5. Halley's Comet had two distinct tails with different curvature and spectrum. Both tails were situated near the plane of the orbit of the nucleus. The average value of $(1 - \mu)$ in Tail I, derived from its curvature, was from 10 to 20. Condensations in this tail indicated far greater repulsive forces. Tail II was of a diffuse character and showed $(1 - \mu)$ from 0.1 to 0.3.

6. The motion of matter in the fine streamers of the tail gave evidence of $(1 - \mu)$ serveral hundred times gravitation.

7. The straight jets in the prolonged radius vector near the nucleus showed small repulsive forces, $(1 - \mu)$ varying from 1 to 6.

8. Motion of matter in the envelopes showed $(1 - \mu) = 1$, indicating approximate balance between attractive and repulsive forces.

9. The nucleus and the neighboring parts of the coma showed in their spectrum the cyanogen, Swan, and Raffety bands, and the D lines of sodium. Tail I in general and the condensations moving along it had the spectrum of CO^+. The fine streamers associated with this tail probably also consisted of CO^+. The jets showed in their spectra a preponderance of CN. The envelopes gave the spectrum of CN and Swan. Tail II consisted of dust reflecting solar light with appreciably the same wave-length of the maximum of intensity.

10. In spite of the difference in the constitution and in the repulsive forces in these various formations they were not independent of each other. The tips of the jets and envelopes generated the fine streamers of the tail, while the streamers coalescing produced the condensations.

11. The nucleus exerted appreciable repulsive action on the neighboring particles. This follows from the accelerated expansion of the halos and the accelerated motion of the jets toward the Sun. A certain amount of interaction was observed between streamers, jets, and other formations.

12. The repulsive action of the nucleus justifies Bessel's idea of the sphere of action of the nucleus and shows that the orbit of a condensation moving in the tail need not intersect the orbit of the nucleus. Consequently all orbits based on this assumption are substantially incorrect.

13. The mass of the nucleus was estimated to be of the order of 10^{-10} times the mass of the Earth.

14. No evidence for the rotation or periodic oscillation of the nucelus was found.

15. The rate of expansion of the spherical halos was from one tenth to serveral km/sec.

16. On many dates secondary nuclei were found showing all the properties of the primary nucleus, namely, halos, jets, and streamers.

17. Processes of an explosive character in the nucleus, head, and condensations of Tail I were observed.

18. The figure of the head departed very considerably from a paraboloid of revolution approaching a figure produced by a catenary.

19. The nucleus, the head, and the tail showed variation in size attaining their minimum (nucleus and head) and maximum (tail) values five to six weeks after perihelion.

20. The existence of carbon isotopes in comets Halley, Brooks (1911 V), and Delavan (1914 V) was established.

21. The cyanogen bands in Halley's as well as in some other comets showed a large displacement to the violet which is certainly not a Doppler effect. The rotational structure of these bands indicates excitation by fluorescence.

22. The deviations in the structure of the Swan bands from the laboratory sources were studied.

23. A number of bands and lines unidentified as yet were found. In most cases they coincide with the secondary maxima in the solar spectrum. Of particular interest is the possible presence of the fluorescence bands of Na_2.

24. Bessel-Bredichin's mechanical theory was found sufficient to explain the observed facts. However, Brédichin's classification of cometary tails was not sustained.

25. Dark formations were noticed near the nucleus. These would be invisible outside the coma. On serveral occasions the existence of secondary nuclei was suspected only from the configurations of the jets. Thus some comentary formations can be entirely missed both visually and photographically.

26. While the primary process is the ejection and transference of matter, in some cases rapid changes in the luminosity of the streamers and jets were noticed. This shows that a transference of luminosity may also play a part in the activity of comets.

April, 1930, University of California, BERKELEY, CALIFORNIA.

N. T. Bobrovnikoff, Publ. of the Lick Observ., Vol. X VII, pt. 2, 1931.

Subject Index

acceleration(s) 55, 59, 69, 101, 144
accompanying star 16
asymmetries 14

brightness variations 13, 15, 17, 18, 19, 20,23
bulb-like shape 21

coma 12, 14, 17, 21, 37, 65, 68
condensations, motion of — 21, 114

dark cavity (space) 46, 63, 67
dark lane 46, 52, 64, 66, 101, 117
discovery 12, 13

ejections 49, 59, 69, 70, 112, 123
elongations of the nucleus 16, 24, 31, 52, 64, 71, 105
envelopes 17, 21, 24, 29, 34, 47, 50, 55, 59, 66, 69, 72, 101, 150
explosions 31, 33, 69, 108

fan(s) 20, 29, 40, 46, 59, 69, 109, 117
filaments 50
first observation 12

globular formations 121
granulation of the coma 16, 17

halo 18, 29, 34, 40, 44, 48, 51, 55, 111, 114, 116, 121, 133, 134, 138, 142, 154

jets 20, 34, 45, 54, 69, 101, 108, 114, 121, 132, 135, 143

motion in fine streamers 44
motions in the tail 21, 104, 144, 146
motions of condensations 30, 32

nucleus flashing 22

observations at Kodaikanal 78
observations at Windsor 152
observations in 1910 154
observations in 1911 156
observation made in Jamaica 41
occasional brightening (flashing) 15, 22, 103, 111
oscillations 35, 47, 56, 78

protuberance 36, 53

Radcliffe Observatory observations 23
rapid changes 23, 44, 48, 103
rays 46, 50, 52, 61, 64, 66, 108, 117
red-yellow colour 37, 52, 62, 65, 85, 100, 117, 193
repulsive force 31, 39, 51, 56, 67, 100, 104, 109, 111, 143

secondary and multiple nuclei 22, 29, 33, 41, 44, 58, 69, 105, 108, 117, 125, 134
spectroscopic observations 14, 21, 31, 37, 40, 45, 49, 53, 57, 63, 68, 71, 78, 85, 110, 115, 123, 136, 138, 142, 148, 150
— of streamers 40, 45
star occultation 106, 107
streamer(s) 17, 21, 44, 52, 66, 72, 109, 114, 122, 132, 149

tail branches 24, 30, 33, 44, 49, 61, 72, 152
tail passage of the Earth 76, 87, 89, 94, 97
tail observations 17, 36, 41, 65, 72, 76, 83, 88, 117, 131, 153
transit across the Sun 74, 75, 77, 88, 90, 96
two tails 47, 48

velocity of ejections 119
velocity of expansion 103, 120

Bei Fragen zur Produktsicherheit wenden Sie sich bitte an:
If you have any questions regarding product safety,
please contact:

Walter de Gruyter GmbH
Genthiner Straße 13
10785 Berlin
productsafety@degruyterbrill.com